RECOVERING CARIBBEAN NATURE

The Caribbean is a global biodiversity hotspot; half of its resident bird species are found nowhere else, yet, a quarter are threatened with extinction. Nearly all its native amphibians and reptiles and thousands of plants also are endemic. Yet, less than 1% of the landscape can be considered natural; and apart from reserves, most land is privately owned. Despite the challenges of such habitat fragmentation, the Caribbean's distinctive fauna and flora can be preserved through planning and managing a connected network of sustainable naturalistic landscapes, reserves, parks, and private gardens. This book uniquely provides both a theoretical background and practical applications to restoring nature within the tropical Caribbean. Packed with beautiful color photographs, it offers unifying principles that can be applied across the tropics and synthesizes information on the Caribbean's environmental uniqueness and globally significant biodiversity. It also provides explicit guidance on establishing sustainable and more naturalistic landscapes from large public lands to private yards and gardens.

The book is essential reading for academics and researchers studying the Caribbean environment, resource management professionals, and scientists and educators from nongovernmental organizations who provide programs and advocacy for conservation and regional sustainability. Moreover, it highlights the importance of private lands and gardens, where the greatest gains can be made, and so offers a handbook for knowledgeable private landowners and their professional advisors.

Recovering Caribbean Nature Blurbs

The recovery of nature is one of the most urgent challenges we face, in the Caribbean and around the world. With more than 1 million species threatened with extinction and the specter of climate change and sea level rise upon us the path forward may seem insurmountable. But the good news is that we can all participate in the creation of a new sustainable world, though personal actions in our gardens and landscapes. *Recovering Caribbean Nature* comes at a perfect time and is a welcome new resource packed with detailed guidance on how to personally contribute to the restoration and recovery of nature.

George D. Gann, Executive Director, The Institute for Regional Conservation and Chair Emeritus, Society for Ecological Restoration

Recovering Caribbean Nature is a monumental "how to" restore the Caribbean native flora. Decades of research, island reconnaissance, and visits to gardens, nurseries and still intact and pristine habitats, provide the authors with unique insights into a growing imperative to restore the region's native flora. As a conservation practitioner for BirdLife International, it's given me a refreshing perspective on the Caribbean natural history. I even think it is the foundation for a much-needed Caribbean-wide rewilding movement – one garden at a time.

Ian J. Davidson, Director-Americas, BirdLife International

Jim Kushlan and Kirsten Hines have consolidated knowledge from a range of resources and their own studies over many years to provide Caribbean-specific information on how to protect, restore, and recreate Caribbean nature. I am confident this book will have a tremendous impact on inspiring more people to plant native, thereby providing more high-quality habitats that are needed to support and sustain the incredible diversity of endemic and migratory birds that call the Caribbean home. Present and future generations and locals and visitors alike will enjoy and benefit from a healthier environment flourishing with native plants and abundant birds.

Lisa Sorenson, Executive Director, BirdsCaribbean

Aprendí y disfruté leyendo este libro, *Recovering Caribbean Nature*. Su revisión de la historia ambiental del Caribe y su papel en la conservación de la naturaleza actual proporcionan valiosos ejemplos que pueden ser útiles tanto para un huerto comunitario como para una universidad. Su revisión de la historia ambiental del Caribe y su papel en la conservación de la naturaleza actual proporcionan valiosos ejemplos que pueden ser útiles tanto para un huerto comunitario como para una clase universitaria. *Recovering Caribbean Nature* enfatiza acertadamente el papel que juegan los jardines en la conservación de las aves caribeñas, como hemos descubierto en el Jardín Botánico Nacional de Cuba. Los autores describen las plantas más adecuadas para la recuperación de la naturaleza así como las aves y otros animales que podrían usarlas, información que ahora estará disponible para muchas personas en el Caribe. Proporciona consejos prácticos y reflexiones sobre la restauración de la naturaleza que pueden guiar las acciones de conservación en todo el Caribe. Libros como este pueden ayudar a las islas de las Antillas a ser más resilientes y hermosas y, al mismo tiempo, convertirse en un mejor lugar en el mundo!

Lourdes Mugica Valdés, PhD, Facultad de Biología, Universidad de la Habana

RECOVERING CARIBBEAN NATURE

JAMES A. KUSHLAN
KIRSTEN HINES

Photographs by Kirsten Hines

CRC Press
Taylor & Francis Group
Boca Raton London New York

CRC Press is an imprint of the
Taylor & Francis Group, an **informa** business

BirdsCaribbean

BirdLife
INTERNATIONAL

Photographs by Kirsten Hines

First edition published 2025
by CRC Press
2385 NW Executive Center Drive, Suite 320, Boca Raton FL 33431

and by CRC Press
4 Park Square, Milton Park, Abingdon, Oxon, OX14 4RN

CRC Press is an imprint of Taylor & Francis Group, LLC

ISBN: 978-1-032-48983-4 (hbk)
ISBN: 978-1-032-49978-9 (pbk)
ISBN: 978-1-003-39632-1 (ebk)

DOI: 10.1201/9781003396321

Typeset in StoneSerif
by Apex CoVantage, LLC

Contents

Foreword

I was most fortunate to come to the Caribbean over 35 years ago as a doctoral student to study white-cheeked pintails in The Bahamas. My engagement has only grown since, first leading programs in wetlands and waterbird conservation and currently as the Executive Director of BirdsCaribbean, a vibrant international network of members and partners committed to conserving birds and their habitats in the insular Caribbean. BirdsCaribbean and I, personally, care deeply about conserving the Caribbean's birds. Through our regional projects and programs, we strive to build a region where communities are uplifted and empowered; conservation actions are guided by robust science and data; and nature is loved, respected, and protected as a core value. Protecting and restoring natural habitats and enhancing the native plants that support the Caribbean's birds and other animals are vital to our work. Everyone can help and play a role!

The Caribbean is recognized globally as one of the world's top five biodiversity hotspots, owing to its many endemic species (including 180 endemic birds) and its history of habitat loss. That such biological richness is contained within a comparatively small land area, an area also supporting large numbers of people, has put many of these natural treasures at risk. Caribbean nature has historically endured landscape-wide impacts owing to centuries of island-clearing agriculture and more recently rampant development, especially for tourism. Yet, Caribbean nature continues to show its resiliency for recovery. Clear-cut forests that were turned into plantations can recover native vegetation; and endangered plants and animals have persisted despite numerous threats, such as overharvest of lignum vitae trees, parrots, hutias, and rock iguanas. This book provides the guidance needed to continue to move forward in recovering Caribbean nature.

I first worked with Jim Kushlan when he was leading the effort to create a hemispheric conservation plan for waterbirds, including Caribbean waterbirds. Jim is also past-president of the American Ornithologist's Union (now American Ornithological Society) and Waterbird Society. Kirsten Hines is a writer and wildlife photographer, having published several books on her photography and writing. Both have broad experience as biologists in the Caribbean. Together, they wrote the definitive book on attracting birds to gardens in South Florida, the northernmost extension of the Caribbean flora and fauna, and another on the roles of a tropical botanic garden in bird conservation. They have spent over 10 years expanding their vision and insights for recovering nature to the Caribbean. This book is the result. In it, they share invaluable history, philosophy, and practicalities of restoring Caribbean nature especially through the use of native plants, providing extensive guidance on plants to consider and the animals that might use them.

Recovering Caribbean nature sounds like something that is done by governments in national parks, and that certainly is so, but, in reality, it can take many forms. It takes the partnerships of government, nonprofit organizations, community groups, and individuals, particularly individual landowners, and managers. Since most of the land in the Caribbean is privately owned, the greatest strides can be made by individuals who choose to set aside portions of the landscape to plant native trees and support native wildlife. This book especially speaks to conservation on private land. Native plants can turn backyards, schoolyards, neighborhood parks and even hotel grounds into natural habitats that attract and sustain birds and other wildlife. In the Caribbean, land is precious – particularly the coastal areas where so many people live. Restoring native habitat within our neighborhoods allows us to share the land with native plants and animals. Of course, it doesn't just benefit the birds. Native landscaping makes neighborhoods more beautiful and islands more sustainable and keeps us in touch with Caribbean heritage and traditions.

BirdsCaribbean has been at the forefront of encouraging conservation of Caribbean birds and their habitats. This highly readable and inspiring book effectively guides us toward these goals. Jim and Kirsten have consolidated knowledge from a range of resources and their own studies over many years to provide Caribbean-specific information on how to protect, restore, and re-create Caribbean nature. There's no wrong way to bring native plants into your backyard gardening or neighborhood landscaping, but there are things to do to make the most of your native plants so as to effectively re-create habitat for birds and other native animals. With the sound and experienced advice provided by Jim and Kirsten in this book, even the smallest yards can play an important ecological role in recovering Caribbean nature.

If you plant it, they will come . . . I am confident this book will have a tremendous impact on inspiring more people to plant native, thereby providing more high-quality habitats that are needed to support and sustain the incredible diversity of endemic and migratory birds that call the Caribbean home. Present and future generations and locals and visitors alike will enjoy and benefit from a healthier environment flourishing with native plants and abundant birds.

Lisa Sorenson, PhD
Executive Director, BirdsCaribbean

Preface

This book emerges from our deep appreciation for Caribbean nature – its birds, reptiles, plants, gardens, landscapes, history, people, and cultures. Our goal is to share information and concepts that may prove useful to those working toward protecting, recovering, and restoring a more-natural and more-sustainable Caribbean landscape. Between us, we have been privileged to experience the Caribbean broadly in nearly all Caribbean nations, conducted ecological studies there, and participated in conservation planning. Our Caribbean is that of endemic birds, seabird colonies, iguana beaches, rocky headlands, forests, gardens, and towns.

Our interest in regreening tropical landscapes evolved in South Florida. Jim has long conducted research studies on birds, reptiles, and fish in South Florida and the Caribbean, set out in hundreds of scientific contributions including over a dozen books. Kirsten studied Caribbean rock iguanas, coauthored the primary online reference for South Florida's native plants, *Natives for Your Neighborhood*, and authored the books *Wild Florida: An Animal Odyssey* and *Birds of Florida*. Nature recovery, regreening, and the use of native plants have a considerable constituency in the United States. In our own conservation work, we found that available philosophy and guidelines for recovering nature did not work very well for the tropical habitats, flora, birds, reptiles, and butterflies of South Florida, mainly because the insights and recommendations that were available were derived from temperate-zone experiences. So, we wrote a book specifically aimed at those working towards recovering South Florida's primarily tropical nature, *Attracting Birds to South Florida Gardens*, and a companion book, *Birds of Fairchild*, on how a public garden can influence plant and animal conservation. South Florida is not the Caribbean, but it is close; in fact, it's the northmost extension of the Caribbean flora.

Attracting Birds to South Florida Gardens is also relevant to nearby limestone islands of the Lucayan Archipelago, Cayman Islands, and parts of Cuba, the origin of most of South Florida's Caribbean plants. This became clear when we considered expanding coverage to Bahama islands and, in the process, became more appreciative of the diversity of the Caribbean as a whole and the nature that is at risk. So, we decided to extend the philosophy, approach, and practical advice by taking a regional perspective. We spent over a decade investigating, studying, and thinking about Caribbean nature. Many of the notes in this book on the interactions of plants and animals stem from our observations. What we learned also is documented in Kirsten's photography illustrating this book. We found that an appreciation of Caribbean nature also necessitated appreciation of its cultural diversity, its history—especially the impact of the region's history of land use—and each island's indigenous biodiversity.

This is a book about the Caribbean's terrestrial environment, its landscape, its indigenous plants, its birds and other indigenous animals, its common species, and its endemics. It is about recovering the environment that supports a global biodiversity hotspot. Conserving indigenous birds and other terrestrial animals of the Caribbean means protecting, restoring, maintaining, and creating habitats for them to use, principally through native plants re-creating native habitat. Recovering nature involves engagement across an island's landscape, its protected areas, public parks, botanical gardens, and especially private land. Birds and butterflies can be sentinels, and we have emphasized them. If they come, they will show that Caribbean nature has been re-created to some degree to their satisfaction. Recovering Caribbean nature is fundamentally about its plants, birds, and other animals and about the people who live with them.

Acknowledgments

We thank many for their advice, guidance, and suggestions over the years of this project and for helping us better know and appreciate the Caribbean: Kevin Achee Lee-Yuk, the late George Angehr, Patricia Bradley, Sandra Buckner, Dale Browne, Wayne Burke, Eric Carey, Andrew Dobson, Ethan Fried, Lynn Gape, Bruce Hallett, Marge Hamilton, John Iverson, Brett Jestrow, Lee B. Kass, Edward Massiah, Predensa Moore, Lourdes Mugica Valdés, Clive Petrovic, Tania Piñeiro Cordero, Joseph Prosper, Bruce Purdy, Lisa Sorenson, Kathleen Sullivan-Sealey, Ann Sutton, Portia Sweeting, Herb Raffaele, Ernesto Reyes, Frank Wadsworth, Carolyn Wardle, David Wege, the late Tony White, the late James Wiley, David Wingate, Natasha Wright, and Joseph Wunderle. We thank and acknowledge Eric Carey, Pericles Mailles, and Joseph Wunderle for encouraging the start of a prior Bahamas-oriented project. We thank colleagues who generously and knowledgably reviewed various manuscript drafts: Eric Carey, Ian Davidson, Javier Francisco-Ortega, George Gann, Lynn Gape, Brett Jestrow, Ariel Lugo, Lourdes Mugica Valdés, Lisa Sorenson, Kathleen Sullivan-Sealey, Joseph Wunderle, and Mark Yokoyama. Their inputs were most helpful and deeply appreciated. At Taylor & Francis, CRC Press, we thank our editorial team Amelia Bashford, Shikha Garg, Marsha Hecht, Balaji Karuppanan, and Alice Oven. All photographs are by Kirsten Hines.

About the authors

James A. Kushlan is a biologist and writer who lives in Coconut Grove, Florida, and Annapolis, Maryland, the United States. In his 12 books and over 250 technical articles, he shares with a diverse audience his perspectives on science, conservation, and natural history. His books include *The Herons; Heron Conservation; Storks, Ibis, and Spoonbills of the World; Attracting Birds to South Florida Gardens; Seeking the American Tropics: South Florida's Early Naturalists* and 3 books on South Florida's national parks. He holds a PhD from the University of Miami and 2 honorary doctor of science degrees. His professional positions have included Director of Patuxent Wildlife Research Center, Research Associate of the Smithsonian Institution, and Professor of Biology at Texas A&M-Commerce and the University of Mississippi. He was the 51st president of the American Ornithologists' Union and president of the Waterbird Society. He has served on many professional and conservation boards, including the American Bird Conservancy, North American Bird Conservation Initiative, Waterbird Conservation for the Americas, as the chair, Wetlands International, John Cabot University, Station Biologique de la Tour du Valat, Zoo Miami Foundation, as the chair, and the Everglades Foundation. The University of Miami has established the Chair of Waterbird Biology and Conservation in his name. More background can be found at www.JamesAKushlan.com.

Kirsten Hines is a Florida-based author, wildlife photographer, and conservationist with a master's degree in biology and background as an environmental educator. Much of her early career research as a biologist focused on endangered rock iguanas in the Exuma Islands of The Bahamas and she continues to serve on the IUCN-Iguana Specialist Group. Her writing and photography have appeared in numerous publications including 8 of her own books, such as the award-winning titles *Attracting Birds to South Florida Gardens, Birds of Fairchild, Wild Florida: An Animal Odyssey*, and most recently *Birds of Florida: A Photographic Guide*. Kirsten aims to inspire conservation action through her storytelling, wildlife-oriented presentations, guiding, and workshops and through not-for-profit work such as being a Conservation Associate at the Institute for Regional Conservation, cofounding Phoebes Birding to connect women through nature, and serving on such boards as Tropical Audubon Society and Audubon Florida, where she is Conservation Chair. More about Kirsten and her work can be found at www.KirstenHines.com.

Introduction

The islands of the Caribbean are places of impressive trees, beautiful flowers, tropical fruits, sunshine, afternoon rain showers, dependable breezes, palm-lined beaches, rocky shores, towering volcanic mountains, the 17th largest island in the world, and tiny cays barely peaking above sea level. Home to over 40 million people and a destination for more than half that many visitors in some years, the 1600–4000 islands of the Caribbean (the number depending on who is counting what rock) are expected to be tropical, breezy, green bits of paradise.[1, 2] Flower-bedecked tropical trees line roads; colorful birds and butterflies flit through shrubs; a day shift of sunning lizards transitions to a night shift of peeping frogs, making nature obvious even to a short-term visitor or cruiser. Tropical plants and animals are a highlight of the Caribbean.

But Caribbean nature needs regreening. How can such an inherently green place possibly need regreening? It turns out that much of its current, obvious green is not due primarily to indigenously Caribbean plants. Bananas, bougainvillea, coconut palms, flamboyant trees, and mangos represent Caribbean scenes that visitors and, indeed, island residents expect – yet these plants are not native to the Caribbean but imported from elsewhere in the world's tropics. The apparent green of today's Caribbean nature is but a shadow of the diverse nature that existed before humans made their first footprints on its sandy beaches. It is similarly a shadow of the diverse nature of indigenous plants, animals, and people that coexisted for millennia prior to Europeans' arrival. In common with most of the world, Caribbean islands are now cultural landscapes built upon and in large measure displacing an island's natural patrimony of indigenous plants and animals. It need not stay that way. Indigenous nature and humanity's relationship to it can be recovered.

For decades, much concern has been raised by government and civil society about conserving Caribbean biodiversity. Depending on political affiliations, Caribbean nations and dependent territories have participated as they can in international biodiversity programs including the UN Caribbean Environment Programme, UN Convention on Biological Diversity, Commission on Sustainable Development, and Small Island Developing States initiatives.[3–6] Nearly all Caribbean nations have formally adopted sustainable development and conserving biodiversity as priorities.[7, 8] And along with over a hundred countries, Caribbean nations have pledged to protect at least 30% of their land surface by the year 2030 and to have restoration completed or underway on at least 30% of degraded ecosystems.[9–13] The decade 2020–2030 is the UN Decade on Ecosystem Restoration, and the task of recovering nature is central to the Caribbean's official approach to biodiversity conservation and sustainability.

DOI: 10.1201/9781003396321-1

A cruising anchorage surrounded by coastal scrub forest in the Grenadines.

In addition to committing to international agreements for biodiversity protection, sustainable development, and ecosystem restoration, governments have done their part where financially and politically feasible by setting aside protected areas and by creating agencies, policies, and programs to protect biodiversity, enacting environmental regulations, taking advantage of multinational funding opportunities, and enabling nongovernmental conservation action. Nationally, these actions create frameworks for protecting and managing indigenous biodiversity. Large governmentally protected reserves form the backbone for recovering Caribbean nature along with an array of smaller parklands and public gardens. Yet, much of the land belongs not to government but to private individuals, families, co-operatives, corporations, and nongovernmental organizations. This private landscape retains ecological potential that can be harnessed by those responsible for a property – owners, tenants, and managers. Such engagement complements government-run programs, usually with fewer constraints. Through combined government and civil society action, much of Caribbean nature still can be recovered.

What is the nature to be recovered? Fundamentally, it is the region's indigenous, endemic, and endangered plants and animals and the natural ecosystems, habitats, and communities they form. As discussed further in Chapter 1, the extent of the Caribbean's endemic terrestrial flora and fauna is significant. Yet, thousands of Caribbean species are known to be threatened with extinction.[14, 15] In Cuba alone, over 2500 species of plants, 46% of the national flora, are threatened.[16] Despite current threats on top of centuries of profound environmental alteration at the hand of humanity, the potential remains for restoring more natural landscapes, parks, and private gardens, thereby enhancing environmental and societal resilience.

Concepts such as recovering nature, ecosystem restoration, regreening, natural landscaping, and environment-sensitive landscaping, and natural gardening have gained traction in temperate zones. But beyond some generalities, temperate experiences translate poorly to the tropics. Tropically adapted floras and faunas, hyper-local biodiversity, limited temperature seasonality, and profound rainfall seasonality of Caribbean islands preclude a temperate approach. The aim of this book is to synthesize and share ideas, insights, and options for recovering nature on public and private lands in ways appropriate to the Caribbean. We hope it might be of value to conservation, restoration, and resource management professionals and be accessible to those many more individuals who also share responsibility or interest in improving the environment, especially through native plants benefitting birds and other animals.

Cuban tody, a tiny endemic bird, one of 5 species of a family of birds that today occurs only in the Greater Antilles.

The organization of the book is first to share a few considerations of what Caribbean nature means and what recovering that nature might entail in large public protected areas and public gardens, communities as a whole, and also on smaller private lands – individual yards, gardens, and similar small holdings. These ideas have general application through the region providing a common point of reference for diverse restorations. Then more practical matters are discussed followed by chapters that share information on ecologically desirable plants and the animals that could contribute to and benefit from nature recovery efforts. These are followed by some concluding thoughts and a guide to additional resources.

Given the natural, cultural, and political diversity among and within Caribbean islands, this book is regional rather than island-specific in scope. To complement such regional background, local experience is essential to account for local distinctions in soils, rainfall, climate, what plants work or do not, what birds and other wildlife might occur, and what cultural and governmental constraints and encouragements there might be. Local knowledge especially is needed to identify what local plants or animals might be of special conservation concern to be used in restorations and natural gardens.

The concept of the Caribbean enjoys multiple meanings. The Caribbean of this book incorporates the insular Caribbean of the Greater Antilles, Cayman and Virgin Islands,

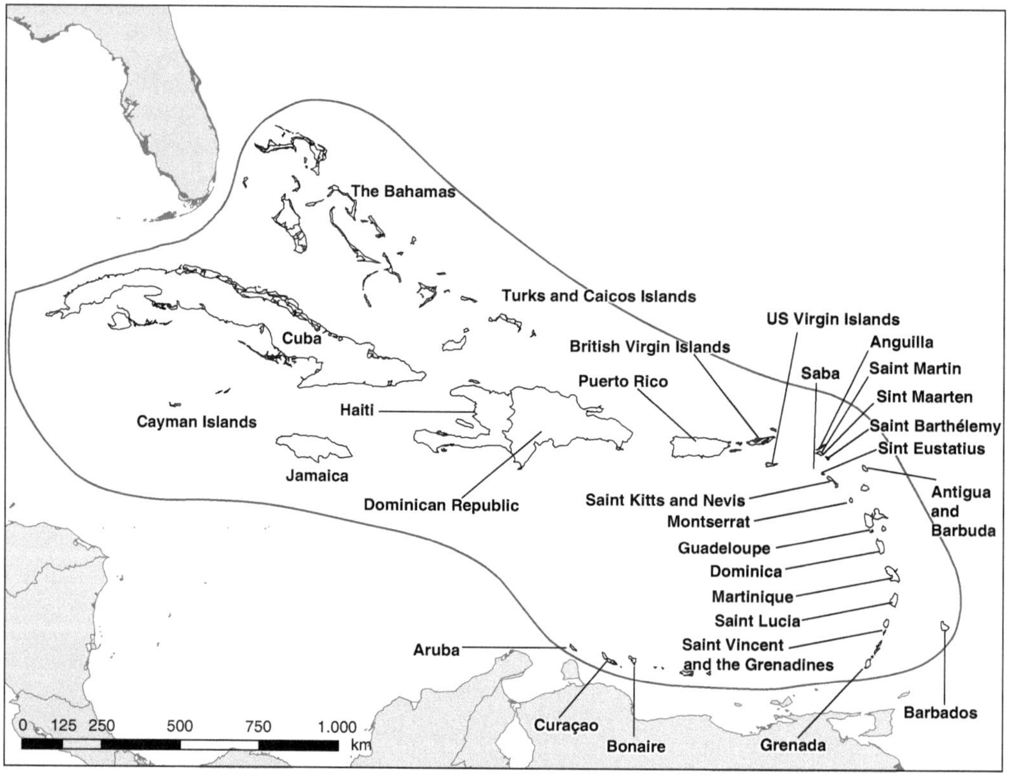

Map of the Caribbean outlining the area covered in this book.

Lesser Antilles, Leeward Antilles (southern Dutch Caribbean), and the Archipelago of San Andrés y Providencia. It also includes Barbados and the Lucayan Archipelago (islands of The Bahamas and Turks and Caicos), which lay in the nearby western Atlantic but are allied biogeographically and considered Caribbean by international organizations. This is essentially the area of Bond's Line illustrating his theory of Caribbean bird biography, with the addition of the Leeward Antilles.[17] The non-continental islands of the Caribbean covered in the book share flora, fauna, habitats, climate, history, and culture. These approaches and information also have general applicability to the wider Caribbean of adjacent continental coasts and islands and to its northern extension in southern Florida, in tandem with our companion book on that area.[18]

The environmental, cultural, conservation, and political diversity of the Caribbean incorporates thousands of islands, diverse cultures, 6 official languages, and 27 sovereign nations and dependent territories. The Caribbean's environmental complexities include its geology largely shaped by the tectonics of the Caribbean Plate, wet–dry climate, hurricane history, soils, topography, and its plants and animals. Limestone islands differ from volcanic ones; mountains from sea level; leeward sides from windward; wet from dry; rocky from sandy; north from south. Cultural complexities include the history of immigrations of Indigenous Peoples, European colonialists, Enslaved People, and later other diverse immigrants each of which added their own traditions to the Caribbean cultural mix. Conservation and political complexities include recovering from the impacts of colonial agriculture,

Cactus patches along the Nevis coast surrounding the ruins of a lime kiln, a remnant of the island's plantation-era heritage.

effects of historical and present land-use policies, diverse national environmental laws and their enforcement, and evolving regional and national political priorities. Each complexity, natural and human, affects what recovery of nature can be accomplished.

The overall goal of the book is to encourage and, perhaps in some small way enable, creation of island-wide landscapes made up of patches of habitat that cumulatively are recovering some semblance of the natural environment. A degraded forest can turn into a reserve, a neglected park into recovered habitat, a lot into a garden, a home yard into a wildlife refuge, bare ground into forest, a commercial lot into a native plant showpiece, a weedy roadside into a linear natural garden, sterile dirt into productive soil – each recovered patch contributing to improving the ecological health of the community, island, nation, and region. Each recovered patch becomes part of a matrix of recovered nature that increases an island's sustainability, conserves water, protects water quality, enhances air quality, reduces pollution, sequesters carbon, and provides many other human and community values. This book is meant as an invitation to engage in a patch-by-patch recovery of Caribbean nature.

References

1. Srinivasan, K., I. Otker, U. Ramakrishman, and T. S. C. Alleyne. 2017. "Caribbean Tourism in the Global Marketplace: Trends, Drivers, and Challenges." In *Unleashing Growth and Strengthening Resilience in the Caribbean*. Washington: IMF. www.elibrary.imf.org/display/book/9781484315194/ch03.xml?language=en&redirect=true

2. Worldometer. 2023. "Caribbean Population." www.worldometers.info/world-population/caribbean-population/
3. UNEP. date unknown. "Small Island Developing States." United Nations Environmental Programme. www.unep.org/explore-topics/oceans-seas/what-we-do/working-regional-seas/small-island-developing-states
4. GLISPA. date unknown. "Global Island Partnership." www.glispa.org/
5. CBD. date unknown. "The Convention on Biological Diversity." www.cbd.int/convention/
6. United Nations. date unknown. "Small Island Developing States." www.un.org/ohrlls/content/small-island-developing-states
7. Bovarnick, A., F. Alpizar, and C. Schnell. 2010. *The Importance of Biodiversity and Ecosystems in Economic Growth and Equity in Latin America and the Caribbean: An Economic Valuation of Ecosystems*. New York: United Nations Development Programme.
8. CARICOM. 2018. *The State of Biodiversity in the Caribbean Community: A Review of Progress Towards the Aichi Biodiversity Targets*. Georgetown, Grenada: Caribbean Community Secretariat.
9. Alvarez Malvido, M., C. Lazaro, X. De Lamo, D. Juffe-Bignoli, R. Cao, P. Bueno, C. Sofrony, C. Maretti, and F. Guerra, eds. 2021. "Latin America and the Caribbean Protected Planet Report 2020." Red-Parques, https://drive.google.com/file/d/1R4AbdemAH7mUgACuXJ0ovqncuJ8no4AU/vie
10. Convention on Biological Diversity. 2023. "Kunming-Montreal Global Biodiversity Framework, 2030 Targets." www.cbd.int/gbf/targets/
11. High Ambition Coalition for Nature and People. 2022. "30×30 High Ambition Coalition Ready to Deliver." www.hacfornatureandpeople.org/home
12. UNEP. 2022. "COP15: Nations Adopt Four Goals, 23 Targets for 2030 in Landmark Biodiversity Agreement." www.cbd.int/article/cop15-cbd-press-release-final-19dec2022
13. United Nations Environment Programme. date unknown. "About the UN Decade." www.decadeonrestoration.org/about-un-decade
14. Mittermeier, R. A., P. R. Gil, M. Hoffman, J. Pilgrim, T. Brooks, C. G. Mittermeier, J. Lamoreux, and G. A. B. da Fonseca. 2005. *Hotspots Revisited: Earth's Biologically Richest and Most Endangered Terrestrial Ecoregions*. Chicago: University of Chicago Press.
15. CANARI. 2019. "The Caribbean Islands Biodiversity Hotspot." Caribbean Natural Resources Institute. www.cepf.net/sites/default/files/cepf-caribbean-islands-ecosystem-profile-summary-2020-english.pdf
16. González Torres, L. R., A. Palmarola, L. González Oliva, E. R. Becker, E. Testé, and D. Barrios. 2016. "Lista Roja de la Flora de Cuba." *Bissea* 10 (Número Especial).
17. Bond, J. 1971. *Birds of the West Indies,* 2nd edition. New York: Houghton Mifflin Company.
18. Kushlan, J. A., and K. Hines. 2014. *Attracting Birds to South Florida Gardens*. Gainesville: University Press of Florida.

1 Caribbean nature

1.1 Land and sea

This book is about the Caribbean's terrestrial environment, recognizing that of the Caribbean's 2.7 million square kilometers, fewer than 10% are terrestrial; the rest are marine. Offshore are seagrass beds, coral reefs, sandy shallows, or precipitous slopes to deep ocean basins and trenches. The biology, conservation, and sustainable management of the Caribbean's marine environment garner considerable attention.[1–4] Although managed through national economic zones, conservation in the Caribbean marine environment is inherently international in scope.

The marine environment is inhabited mostly by species held in common throughout the Caribbean basin. As a result, islands support few endemic marine species. Recognizing the quite-appropriate attention given to protecting the Caribbean marine environment, saving coral reefs, and managing fisheries, it is the Caribbean terrestrial environment where historically and today indigenous (native) and endemic plant and animal species are most at risk. It is the terrestrial environment that merited the Caribbean's recognition as a global biodiversity hotspot. This book focuses on how we might recover nature on Caribbean land.

1.2 Caribbean biodiversity

The importance of the Caribbean's biodiversity is well recognized and documented.[5–15] Its place as the world's third most important biodiversity hotspot is based on its endemism and history of habitat loss.[5] Caribbean islands have about 11,000 species of indigenous plants, a rather remarkable concentration of 3% of the world's known plant species on 0.15% of the world's land area. The indigenous fauna includes about 100 species of mammals, 600 reptiles, and 200 amphibians. Over 700 bird species occur in the Caribbean, year-round or as seasonal residents, with over half of these in Cuba. The Caribbean lies on the flyways of nearly 120 species of birds that breed in North America but migrate to the Caribbean to spend the northern winter or pass through heading further west or south and then return northward in spring.[15–19] Some species that nest in the Caribbean also migrate further south in winter. Neotropical migrants may spend more of their year in the Caribbean than in North America. These part-time residents and passage migrants are as much a part of the Caribbean's biodiversity as are its year-round residents.

DOI: 10.1201/9781003396321-2

It is not the total number of species that makes the Caribbean so biodiverse. Each island actually has fewer native species than an equivalent area of nearby continent, owing to limited space and habitat diversity, and isolation. Using birds as an example, Trinidad, once connected to South America, has twice the bird species as nearby Puerto Rico, an island twice Trinidad's size. Next door, continental Venezuela has over 1,300 bird species, twice that of the Caribbean as a whole. Rather, the key to Caribbean biodiversity is endemism, species occurring nowhere else. The Caribbean tallies an impressive list of such endemic species, and modern genetic techniques continue to reveal previously unrecognized endemic species.

Endemism in Caribbean plants is impressive.[8–10, 20–21] Almost 7,900 (72%) plant species are endemic, as are approximately 200 genera (the next more encompassing taxonomic classification). Moreover, half of these unique plant genera are represented by a single species, and half are found on single islands. Much of this remarkable botanical distinctiveness is concentrated in the Greater Antilles, a consequence of the islands' size, habitat diversity, long isolation, and having been available for plants and animals to colonize for 40 or so million years.[22] Over 50% of Cuba's plants and over 40% of Hispaniola's are endemic. In contrast, on the geologically younger Bahamas, about 10% of the indigenous plants are endemic.

Western spindalis, an endemic species representing a family of birds found nowhere else but in the Caribbean. Long thought to be a single widespread Caribbean bird, modern taxonomic studies have identified that there actually are 4 island-specific endemic species of spindalis in the Greater Antilles.

Almost as remarkable as plants, about 180 species of birds are endemic to the Caribbean, most confined to their single island.[11] It might be expected that birds once reaching islands would then move among them; but this did not happen for many species. As with plants, isolation has been so complete and evolution underway for such a long time that 26 genera and 7 families (next taxonomic grouping above genus) of birds are found nowhere else.[15] BirdLife International has identified 8 areas of exceptional occurrence of endemic birds (called Endemic Bird Areas) within the Caribbean.[23]

For native Caribbean amphibians, endemism is total; all 200 species of native Caribbean amphibians are endemic, most to a single island. Whistling frogs alone (*Eleutherodactylus*) have radiated into 150 species. The Caribbean frog fauna includes the third smallest frog in the world, the Monte Iberia dwarf frog (*Eleutherodactylus iberia*) from Cuba, and one of the largest, the giant ditch frog weighing in at a kilogram (*Leptodactylus fallax*) from Dominica and Montserrat.

Reptiles are nearly as diverse as amphibians. Of the 600 native species, 82% are endemic – 2 groups particularly, the nearly 170 species of tree lizards (*Anolis*) and the 85 species of dwarf geckos (or sphaeros) (Sphaerodactylidae), such as the tiny Virgin Islands dwarf sphaero (*Sphaerodactylus parthenopion*).[10, 22] Snake diversity is concentrated in 2 groups, boas (Boidae) and fossorial blind snakes (Scolecophidia) with over 60 species between them.

Half of all native mammals are endemic. Thirty-five of the nearly 60 species of bats in the Caribbean are endemic, despite their ability to fly.[24] There is an endemic rodent family, the hutias (Capromyidae), as well as an endemic family of shrew-like insectivores, solenodons (Solenodontidae).

Because of its many species, endemic species, and threatened species, 324 places in the Caribbean have been identified as Key Biodiversity Areas for conservation (called KBAs).[11] For birds alone, over 280 Important Bird Areas (called IBAs) are recognized in the Caribbean, covering 16.5% of the region's land area, the largest percentage of areal coverage identified as crucial for bird conservation of any region of the world.[25] Cuba alone has 28 locations meriting recognition as Important Bird Areas.

1.3 Accumulating biodiversity

How did the Caribbean accumulate such biodiversity? Evidence suggests that most ancestors of native Caribbean terrestrial species originated in South America, fewer in Central America, and a few in North America. Islands closer to the continents share more species with their source than islands further away. How plants and non-flying animals got to the Caribbean has been a matter of consideration for decades. Caribbean geologic and biogeographic histories remain active fields of study today.

The islands of the Lesser and Greater Antilles originated through the movement, collisions, uplifts, and subductions of tectonic plates. The Greater Antilles are geologically complex, developing over multiple geologic eras by diverse processes. The Lesser Antilles is of direct volcanic origin sitting on the edge of the still-moving Caribbean tectonic plate. The Lucayan Archipelago is sedimentary limestone with current islands dotting large shallow banks that changing sea levels periodically inundated and exposed. It is generally held that none of the Caribbean islands were attached to continents in any time frame to have major influences on the origin of contemporary species. However, it is possible that the Greater Antilles was briefly (in a geologic time sense) part of a land bridge arc about 35 million years ago when sea level dropped significantly.[26, 27]

The coming and going of Caribbean land over millions of years determined what islands were available for colonization by plants and animals at any point in time, which could successfully colonize the habitats that were available, and which could survive changing terrestrial environmental conditions including those created by sea-level fluctuations. Biogeographically speaking, the older, larger, higher, less isolated, and more physically and geologically diverse an island, the more native and endemic species it will have accumulated over time. Species arrived slowly from continental sources; and, within the Caribbean, the old, large, tall, diverse islands of the Greater Antilles provided colonists for newer islands. Cuba, occupying one-third of the Caribbean's terrestrial area, is the ancient biogeographic heart of the Caribbean.

To reach land not connected to the continent or to other islands, terrestrial plant and animal colonists flew, floated, rafted, or otherwise dispersed over thousands and millions of years in a slow, unpredictable, serendipitous process. Birds, bats, fern spores, some insects, wind-dispersed seeds, spiders, and other small invertebrates were able to fly or were small enough to stay airborne while being carried along by storms or seasonal winds. Seeds floated; crustaceans, reptiles, and some mammals could have been borne by ocean currents; salt-water-resistant reptiles, aestivating snails, and perhaps frog eggs could have come on vegetation rafts. Seeds were transported in the stomachs or on feathers of birds,

Allen Cays rock iguana, a large grazing lizard endemic to a few small Bahamian islands, 1 of 10 extant species of rock iguanas endemic to the Caribbean.

the reason a large portion of the trees of the Caribbean have fruit eaten by birds. Beyond island-hopping, if a short-lived late Eocene land bridge did exist, it would have provided a second dispersal path for ancient lineages of groups such as rodents, sloths, and some invertebrates.[25, 26]

Successful colonization over water would have been difficult, and for the immigrating animal, this is totally unexpected. Most potential colonization attempts surely failed as transport was perilous; arrivals were too few to create a sustainable population; they failed to encounter the right food or habitat; they were eaten by novel predators they encountered; they found their niches already securely occupied; or they otherwise were unlucky. Those that persisted could evolve in isolation from related species and respond to open niches or lack of competitors and predators. Presented with novel opportunities, island species took the ecological place of species that did not make it. Land crabs took on the ecological roles of terrestrial omnivores. Birds, including flightless species, and lizards took the place of carnivoran mammals; lizards claimed roles of insectivorous birds; tortoises, roles of large grazers. Some groups of species diversified widely into different niches. And single species of tree lizards diversified even within an island. Some species are so highly successful in their adopted niches that their populations are much more abundant than they would be on a similar patch of continent land. Each island has its own geologic, biogeographic, and evolutionary history influencing its accumulation of biodiversity.

1.4 Biodiversity and environmental change

Environmental change has been an ongoing process on Caribbean islands since each emerged from the ocean. Once a colonizing plant or animal developed a viable population, it had to persist in face of subsequent environmental changes such as long periods of dry or wet climates, plant community shifts, sea-level fluctuations, volcanoes, land movements, and hurricanes. Not counting current climate change, the last great climatic shifts were at the end of the Pleistocene when deglaciation raised sea levels, covered shallow banks, decreased island size, and provided a wetter climate, creating the Caribbean of today. This was the relatively stable environment encountered as people began emigrating from South America.

Indigenous Peoples began those emigrations about 6,000–7,000 years ago. Human dispersal through the islands continued through several waves by different groups through the millennia, creating the mix of indigenous cultures in place at the time of European contact.[28–36] Fossil records and other documentation suggest that these people found their new lands about 90% covered with forest and diversely inhabited by plants and animals. Some of these plants would have been familiar as these species or near relatives also inhabited the nearby continental mainland. Depending on combinations of location, geology, rainfall, and altitude, immigrating peoples found indigenous plants creating communities such as forest, coastal scrub, mangrove swamp, grassland, freshwater marsh, and arid areas. Some of what they found were extraordinary – 200-kg rodents, giant tortoises, flightless owls, and trees large enough to be fabricated into the 15 m-long canoes noted by early Spanish chroniclers.

Environmental alteration started early; and, as is the case everywhere in the world, the Indigenous Peoples of the Caribbean cleared land for housing and farming, burned forests, felled timber, hunted, fished, and brought with them some of their customary plants and animals. Human hunters harvested the wildlife, some of which were extirpated. Prior to human arrival, the Caribbean, like the rest of the world, had undergone postglacial climate change, which also had caused species extinctions. But we now know that 60% of post-Pleistocene mammal extinctions were due to humans.[26, 34] Disappeared native species include rice rats (Oryzomyini), a falcon, hummingbirds, hutias, macaws, monkeys, owls, parrots, rails, and sloths.[36]

When colonizing Europeans arrived in the Caribbean, the islands were well populated with people, still covered in forests, supported unique birds and other animals, and were surrounded by productive seas. Indigenous People had come to coexist in sustainable ways with the remaining flora and fauna and with the landscapes that had developed over the millennia of their presence. At the time of European contact, the 2 most widespread indigenous cultures were island-traveling mariners, farmers, fishers, and hunters living in chiefdoms and expanding their occupancy of islands. Other indigenous cultures were more localized. Drawing from millennia of experience, these Indigenous People must have had a deep knowledge of their resource-providing environment. Many extant words describing Caribbean nature are derived from Taíno passed through Spanish – ameiva, anoli, boniata, higuana, huracan, mahiz, mamey, manatee, papaya, tabaco, and yuca are readily recognizable. The natural history knowledge of the Caribbean's Indigenous Peoples persists within modern cultures and traditions.[30]

Feral kid goat hiding in the bush on Bonaire. Goats, although important to the food supply and culturally, are among the grazing stock that can quickly devastate native plant populations and with time similarly devastate natural habitats.

Europeans were keen to know about Caribbean nature in that natural and agricultural products were among initial motivations for colonization; and agriculture continued to be at the center of colonial interests for the centuries to follow. As early as the 1500s, European naturalists were sent to document indigenous American flora and fauna. Naturalists and collectors continued to survey the colonies; naturalists and historians back in Europe studied and wrote about the findings; Caribbean plants were cultivated in hothouses and gardens. Much was learned from Indigenous People directly or later through indigenous knowledge preserved by enslaved African people, who continued and evolved Caribbean and African medical and food traditions based primarily on native plants.

The environmental situation changed dramatically during the 500 years of colonial dominance as waves of intensive land use devastated Caribbean nature – clearing, burning, farming, pasturing, lumbering, and plantation- and city-building. Whatever environmental changes had been wrought by Indigenous Peoples paled in contrast to the landscape-clearing of colonists, reducing and mostly eliminating extensive areas of native habitats and their indigenous plants and animals. Colonials and their successors also introduced nonnative plants and animals, including invasive plants such as the African marabú tree, habitat-altering livestock such as European goats and pigs, and mongoose from Asia.

Centuries of reciprocal plant and animal exchange followed, with Europeans enjoying American foods, medicines, and other natural products and colonists trying to make the American tropics look more like the Europe they knew. Breadfruit, citrus, hibiscus, pigeon pea, rice, and turmeric entered the Caribbean landscape and cuisine. Enslaved people brought African plants and traditions for their use, which passed on through the centuries in the Caribbean. Later immigrants brought their own favorite plants. Nuisance weeds hidden in cargo made unplanned exchanges among Europe, Africa, Asia, and the Caribbean. A biogeographic mix of herbs, spices, fruits, ground provisions, and other distinctive food plants, native and exotic, became foundational to Caribbean cultures. Trade crops also became a mix of American and Old-World species – bananas, coconuts, cotton, hemp, lumber, rice, sugarcane, and tobacco. By the modern era, plants growing and being used throughout the Caribbean landscape and the habitats they created were far different than what they were when Europeans arrived.

The current Caribbean landscape is tied tightly to and constrained by the infliction of colonial agriculture, initially driven primarily by the sugarcane plantation economy that dominated the region involving massive land clearing, intense agricultural practices, soil depletion and erosion, and consequent removal of native habitats. By early in the colonial period, except where infeasible such as on steep mountainsides and extremely xeric areas, most islands had their primeval forests, lowlands, and seasonal scrublands cleared, burned, grubbed out, and planted in the monoculture crop of the moment or turned into pasture grass. The impact of the plantation era on natural vegetation and species was profound. On some islands, plant-cover change was nearly complete. Commercial agricultural expansion and contraction repeated themselves in response to changes in the international demand for specific crops. For much of the plantation era and after, small-plot gardens, including those of Enslaved Peoples, provided much of domestic food production. As a result of these histories, some land repeatedly became intensively used and then abandoned to recover naturally.[37]

Beyond farming, colonial-era environmental change involved harvesting the forests. From the early days of Spanish occupation, trees of immediate value were felled and sold. Cedar, lignum vitae, mahogany, mastic, pines, and horseflesh were harvested for their wood; cinnamon, for seasoning; brasiletto, for dye; red mangrove, for tannin; and buttonwood, for charcoal. When alternative fuels were not available, everything that could be burned, by necessity, was burned. Plants were stripped from islands; once abundant species were extirpated or nearly so, surviving only as unsalable specimens.

This assault on the Caribbean landscape is not restricted to long-ago colonial history nor to agriculture but continues today. Forty million residents and half that many tourists need to be accommodated on the region's limited land area. Chain saws replaced cutlasses, and bulldozers replaced grubbing. Rather than small garden patches and home sites being cleared, tracts are leveled for larger-scale development – for residential complexes, commerce, and tourism. Large-scale clearing generally precedes actual development, which often fails to follow, leaving one more patch of land stripped of its natural plant cover. With the coast now being well-claimed by tourism and urban developments, housing and other development creep up the mountains. Environmental change continues.

The exploitation of natural resources and the environmental changes that come with it remain a significant driver of island economies, especially if the role of the environment in attracting tourists is factored in. Beyond millions of tourists and thousands of vans, tourism impacts include ports, hotels and resorts, tourism-industry housing, pollution, trash loads, and consumptive water use leading to aquifer depletion and saltwater intrusion. Export agriculture continues to be important on some islands, although home production for domestic use and sale is often officially encouraged for food security and sustainability. Forestry, including tree plantations, occurs on some islands; and fishing can be economically important for local consumption and for tourism. Mining continues in the Greater Antilles. Hunting, capture of birds and reptiles for the pet trade, and habitat alteration and loss also adversely affect animal species. Banking aside, and tourism included, natural resources underpin the Caribbean economy.

Given the pattern of environmental change, it would be expected that all are not going that well for Caribbean terrestrial biodiversity. And that would be right. Limited ranges and habitats of many species make them vulnerable to catastrophic events such as volcanoes, hurricane wind and surge, disease, land clearing, and epidemics. Other human sources of unnatural mortality include sport and subsistence hunting; mongoose and cats; domestic and export pet trade; nuisance animal control; pesticides; disturbance by there being so many people; and local cultural aversions to certain kinds of animals such as snakes, some lizards, or owls. Of all this, habitat change is the most important cause of extinction or endangerment of Caribbean birds.[15]

As a result of historic and ongoing environmental change, many remaining Caribbean species are threatened with extinction. As of 2010, 55 birds, 26 mammals, 184 reptiles, and 146 amphibians are considered globally threatened.[11] Of native Caribbean amphibian species, 77% are globally threatened, the largest percentage of an amphibian fauna at risk anywhere in the world. An analysis of about 60% of the Cuban flora found over 2,000 species to be globally threatened.[12] Of 35 endemic bat species, 13 are globally threatened because of habitat loss, mostly the loss of old-growth native trees, and the destruction and disturbance of caves. Although analyses of threat have not been done for many

An urban view of Montego Bay, Jamaica.

taxonomic groups over the region, such numbers suggest the situation. Native Caribbean terrestrial biodiversity is in trouble.

Currently, rapid environmental change also is being propelled by ongoing climate change.[38–42] It is expected that these changes will include a rise in sea level by a meter or more by the end of the century, accompanying increase in storm surge penetration, increased saltwater intrusion into porous coastal aquifers, reduction in island size, and stronger, wetter, slower hurricanes putting islands fewer than 3 m above current sea level at risk. Also predicted are longer dry seasons, reduced humidity, higher air temperatures, increased evapotranspiration, reduced rainforest cover, increased dry habitats, lowered freshwater aquifers, and ocean warming and acidification.

Among the mitigation and adaptation measures that have been proposed on island nations is restoring degraded environments. Restoration of mangrove swamps and coastal wetlands is among the top priorities. Restoring degraded terrestrial environments, reforestation, retarding desertification, enhancing water supply management, and instituting more-sustainable land-use practices, especially in agriculture, can ameliorate some impact and slow down the effects of environmental change, allowing restored natural areas more time to accommodate. Recovering aspects of Caribbean nature enhances resiliency.

1.5 The terrestrial environment

Asked to describe the Caribbean's terrestrial environment, a tourist no doubt would comment on its beaches. However, coastal Caribbean also includes rocky shores, mangrove

Diverse habitats on Jamaican mountains.

swamps, and xeric vegetation, perhaps beyond the interests of most visitors. Caribbean landscapes also include pinelands, scrub, rainforests, and high mountain elfin forests. The Caribbean's terrestrial environment, as diverse as its many islands and parts of islands, may be characterized by its vegetation, soils, and climates.

1.5.1 Vegetation

There are many ways to classify Caribbean vegetation. International classification schemes identify 104 vegetation types.[43] Other classifications identify many fewer.[11, 43–47] Technical definitions may be less important than local recognition. International naming schemes notwithstanding, vegetation can be broadly similar on different islands. Generally, Caribbean plant communities can be recognized as variations among forest, scrub, and xeric vegetation but there can be blending between habitats.

Much of the Caribbean is coastal lowland. Vegetation there tends to be similar among islands, as plants can disperse between them colonizing similar environments on different islands. Lowlands are relatively flat to hilly, characterized by a benign, warm, coastal, wet-dry seasonal climate. Lowlands receive upgradient runoff, may have relatively deep if infertile soils, and accumulate water in aquifers, pools, and slow-flowing streams. In contrast, on mountains, particularly above about 1,500 m, conditions are more extreme ranging from wet to dry depending on air currents. They may have steep slopes and valleys, volcanic- or metamorphic-derived soils, cool nights contrasting with hotter days, rapid stream flows, and high insolation or, contrastingly, persistent fog. Diversity and endemism increase on mountains as the plants and animals have had to adapt to the harsher environments that differ along elevational gradients. And unlike the connected coastal lowlands, plants and animals are isolated from other mountain populations leading to endemism.

Overall, it is forests of several sorts that are keys to the region's terrestrial biodiversity. Until the 1500s, the terrestrial Caribbean was predominantly forested, 99% of which were tropical forests.[48] Now overall, about a quarter of the landscape is forested, very broadly defined to include secondary growth of various sorts including tree plantations. Existing forest cover differs among countries, ranging from about 2% in Aruba to about 60% in Dominica. The Greater Antilles, outside of Haiti, have the largest areal extent of forests. Areal coverage of forest – again, broadly defined to include tree plantations – has increased on some islands such as Cuba, the Grenadines, and Puerto Rico as former agricultural and pasture lands regenerate, tree farms are established, and areas are set aside for conservation.[11, 37] Other islands continue to lose forests to development as in the U.S. Virgin Islands or to fuelwood and subsistence agriculture as in Haiti, or for other reasons. Smaller, flatter islands, originally with coastal forests, are the most likely to have lost all of this habitat over time. On mountainous islands, there is a greater chance that high rain and other forests remain. Although such forests on steep mountains and old watershed reserves are quite impressive, most of the current forests turn out to be no older than about a century or so. For the most part, conservation of biodiversity in the Caribbean is a matter of saving, recovering, and connecting wooded patches.

Coniferous pine forest on Abaco in The Bahamas.

Rainforests occur on windward lowlands and lower montane regions, such as in Cuba, Jamaica and Puerto Rico, Hispaniola, and some of the Lesser Antilles. Most remaining nearly intact rainforests are in places historically inaccessible to agriculture or forestry, where colonial-era reserves inhibited harvest, or are secondary growth. Ascending mountains depending on location, rainforests transition to cloud forests and elfin forests, both characterized by their persistent dampness, or to coniferous pine-dominated forest usually considered to be subtropical or even temperate in character. Cloud forests are often enveloped with mist, cutting down sunlight, and are nearly perpetually wet with significant diurnal swings of temperature. Elfin forests are damp and cool, especially at night, with stunted trees covered with a high diversity of mosses, ferns, and bromeliads that can manage the higher UV light of the habitat.

Tropical coniferous forests, called pinelands or pineyards, occur at lower altitudes such as in the Lucayan Archipelago and Cuba, particularly in limestone areas. They can range from being seasonally wet in the lowlands to drier on hills. They also occur at higher elevations, including in Hispaniola and Cuba, where high humidity combined with cool nights inhibits broadleaf trees. Characterized primarily by their tall pine trees, these habitats may have distinctive, diverse, sometimes endemic, low understories of palms, shrubs, and herbaceous plants that may also occur in nearby broadleaf forests. Depending on location, pine forests may be adapted to fires, seasonal flooding, droughts, heat, cold temperatures on mountains, poor-nutrient soils, and even some salinity.

In seasonally drier locations, forests are relatively short, deciduous, or semi-deciduous, and often thorny. They occur throughout the Caribbean and throughout were cleared for colonial agriculture. This structurally diverse plant community has lots of names on different islands such as coppice, hammock, or more colloquially the bush. In this book, we use the term scrub. Trees are a few-to-a-dozen-meters tall interspersed with shrubs and herbaceous plants. There often is not a clear distinction between shrubs and trees, a species may be either. The sorts of plants found in different places depend mostly on the soil and climate, particularly the amount and seasonality of rainfall.

A continuum of increasingly dry conditions produces a continuum of more-and-more desertic habitats. Xeric scrub occurs throughout the Caribbean, particularly on the leeward side of higher islands such as southeastern Cuba and the southern Lucayan Archipelago. Such arid areas are characterized by thorny, deciduous, drought-adapted plants that do not form canopies, and by native cactus, agave, and other succulents, and annuals that come and go with the rains. Trees when present are shrubs. At the dry extreme these habitats are semidesert or cactus scrub, dominated by cactus and lacking ground cover, found particularly in the Leeward Antilles. Xeric scrub also occurs along inland and coastal sand dunes and rocky coastlines, which are physiologically arid environments. Native coastal plants are shrubs and small trees, sometimes wind-shaped, grasses, agaves, and cactus that can accommodate to salty soil and also salt spray, wind, and storms.

Apart from terrestrial environments that are the focus of this book, the Caribbean also has many wetlands including rivers, lakes, blue holes and smaller solution features, ponds, freshwater marshes, mangrove swamps, riparian gallery forests, and freshwater swamps and marshes. In many such habitats, water levels fluctuate seasonally; and in some habitats, fresh water may be perched on denser layers of saltier water or may be connected to the ocean and show tidal fluctuations. Some otherwise terrestrial vegetation, such as low pineyards, also may flood seasonally.

As most of the Caribbean landscape is culturally derived, much of the landscape is far from natural. Farms, plantations, disturbed forests, pastures, ruderal vegetation, yards, villages, cities, ports, roadways, and surface mines are highly altered environments. Many existing savannas and grasslands for the most part also resulted from past human activities. It is these culturally derived environments where efforts at recovering Caribbean nature may have great effect.

Xeric scrub on Bonaire.

Blue hole in the Berry Islands in The Bahamas.

1.5.2 Soil

Understanding the terrestrial environment starts with the soil.[49] The soil found in any patch of an island is influenced by the bedrock underlying it, what has happened to that rock over time geologically and historically, and what has been deposited on top of it. Weathering of rock and rearrangement of soils are both long-term and continuing processes, rocks break up, rivers and hurricanes wash soils down-gradient, mountainsides slide into valleys, and limestone solidifies in real time imbedding coke bottles in beach rock. Human activities such as dumping, farming, digging, grubbing, surface-scraping, excavating, filling, and building all profoundly alter surface rock and its soil. What is found on the surface of any plot is the result of millennia of natural forces and centuries of human impact.

Soil chemistry makes a difference. The soil determines what plants can grow in a location. Where the surface rock is volcanic or metamorphic, soil tends to be acidic. Most of the Caribbean population lives in coastal lowlands, where on all but volcanic islands, soils are primarily limestone, sand, or clay. Where rock is limestone, soil tends to be basic. Sandy soils are more acidic. Sandy and rocky soils tend to have little organic matter so have limited water-holding capacity, and so rain water percolates through and carries nutrients with it. Clay soils slow water infiltration and air penetration, which inhibits plant roots. Most plants do best around neutrality; to grow in acidic or basic soils, many Caribbean plants have developed physiological abilities to deal with one or the other soil extreme but seldom both. Extreme soil conditions limit plant diversity in any one site but provide opportunities for endemism of specially adapted plants. Heavy metals and low fertility of serpentine soils of the Greater Antilles present exceptional challenges to non-adapted plants. Native Caribbean soil is enhanced by atmospheric dust, the major component coming from the Sahara Desert.[50] Most of this clay-sized dust accumulated during the Pleistocene when the African desert and dust storms were more extensive. But dust plumes continue to arrive in the Caribbean in spring and fall. Sahara dust contains silica, aluminum, iron (both soluble and insoluble), phosphorous, calcium, magnesium, and potassium; it accounts for the redness of some Caribbean soils. Red limestone-based soil is prime agricultural land, such as in Cuba.

Over much of the Caribbean, traditional soil hardly exists, at least not in the classical sense shown in drawings of stable, thick, stratified soil horizons in temperate-zone biology books, mainly because humus accumulation tends to be limited. Humus is the partially decomposed product of organic materials littering the soil such as leaves, branches, roots, fruits, and seeds and the once-living components of the detrital food web. In temperate climates, winter interrupts decomposition allowing humus to accumulate, weather slowly, and build soil structure. Generally, in the humid warmth of the tropics, plant debris breaks down quickly owing to the action of fungi, bacteria, termites, worms, and various detritivores. Fire and moisture enhance the oxidation of surficial organic material. As a result, little organic matter accumulates.

The most complete soil development is within forests, especially in stable forests and mountain valleys. There surficial litter layer typically is underlain by a layer of humus followed by mineral layers of iron oxide and clay over bedrock. The usually thin, often-discontinuous litter, and humus layers provide the organic component. However, some upper montane forests may have deep organic humus, especially in cooler habitats such as cloud forests, but the soil tends to be acidic and nutrient deficient. Many tropical trees have shallow root systems and mycorrhizal symbionts that allow access to nutrients near the surface; fewer have tap roots, mostly functioning in anchoring the tree to the ground. In seasonal and scrub forests, soil may be an unconsolidated mixture of litter, humus, inorganic particles, and/or rocks. Sandy soils occur along shores, stretching inland along windward coasts, on fossil dunes, and in inland plains on larger islands. Sandy-soil vegetation ranges from coastal forest and scrub to grasslands. Plants indigenous to these areas are adapted for continuous or seasonally dry conditions. In many places in the Caribbean, especially small rocky islands, crumbled rocks, or the bedrock itself, perhaps with bits of organic soil confined to depressions, are all that pass for soil. Plants there may resemble bonsai.

Human land-use has further compromised Caribbean soil. Historically, seasonally wet–dry lowland scrub vegetation was extensively converted for agriculture and sandy soils converted to pasturage. Large-scale clearing for plantations and other agricultural practices, from the 1500s onward, led to soil erosion and compaction, nutrient leaching, salinization, reduced water availability, and creeping desertification. Despite centuries of use, some soils remain farmable, although generally not extensive on most islands amounting to about 15% of the land in the Dominican Republic and about one-third of Cuba, which for several decades has emphasized restoring agricultural lands using more-sustainable farming practices. Red and limestone-derived soils with organic content can be quite fertile. Most Caribbean agriculture, however, depends heavily on artificial fertilizing and irrigation. Nearly all of Antigua, for example, was covered by agriculture for 300 years, clearly something that would determine the character of current soils.

Stunted blacktorch growing out of bare rocks.

1.5.3 Climate

Nearly all the Caribbean is officially tropical.[51–53] The Tropic of Cancer passes north of Cuba and through The Bahamas. However, it is hard and somewhat misleading to generalize over the entirety of the region. Temperatures do tend to be mild region-wide, averaging about 26°C; but temperature and its variation differ with latitude, altitude, proximity to the ocean, rainfall and humidity, and wind direction. Altitudinally, generally, it is warmer with a daily variation of about a few degrees in the lowlands and with altitude becomes cooler with greater daily variation of about 10 degrees. Higher mountains maintain a more temperate climate. The Köppen–Geiger climate classification system based on plant communities recognizes climate zones in the Caribbean ranging from tropical rainforest to arid, mostly forests of one sort or another.

Latitudinally, temperature trends cooler in the north than in the south because of the effects of the northern hemisphere winter. Seasonally cool temperatures can bring out jackets in Nassau and Havana as cold fronts penetrate southward. These bring cool continental winds from the north or northwest replacing the moderate oceanic easterlies. Continental effects lessen southward in the Caribbean. In Grand Bahama, being convenient to cold fronts, the average low temperature is about 17°C. This contrasts with Dominica at 23°C and Aruba at 26°C. Wind and rain patterns and geography also affect temperature. Quite logically, it is hotter during the day in xeric areas than in wet lowlands; hotter on the dry side of islands than on the windward and in sunny areas than in rainy and cloudy areas. The Caribbean is tropical, but that has somewhat diverse meanings depending on location.

Temperature aside, the Caribbean is actually quite seasonal albeit perhaps a bit more subtly than temperate-zone Autumn leaf-fall or winter blizzards. There are 2 seasons in the Caribbean – the summer-rainy-hurricane season and the winter-dry season. Over

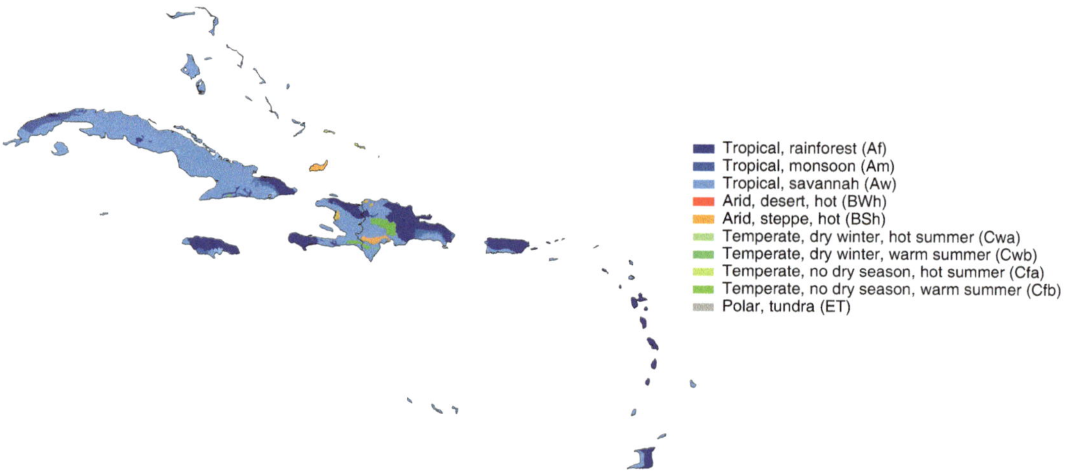

Köppen–Geiger climate classification map for the Caribbean, showing the predominance of tropical monsoon and tropical rainforest climates.

Source: Courtesy: Wikimedia Commons, https://creativecommons.org/licenses/by/4.0/deed.en

Doña Juana Falls, Puerto Rico, prior to a rainstorm.

Doña Juana Falls, Puerto Rico, taken 3.5 h after previous picture following heavy rainfall.

much of the Caribbean about three-fourths of annual precipitation falls in rainy season. For examples, in Nassau an average of 78% of the annual rain falls during May to October and the same from June to November in Dominica. Rain peaks in the early wet season at low elevations and later in the wet season at altitude. For plants and animals, the wet season–dry season cycle is fundamental. Even within a season, variations in rainfall can be important. Heavy rains can rapidly change water conditions experienced by plants and animals.

Every year, the rain cycle is somewhat different as well; some are wetter, some are drier, and this too is not unimportant ecologically as plants and animals need to deal with variation in extremes. Rainfall differences among years occur in part because of cyclical changes in ocean temperature patterns, called the Southern Oscillation in the Pacific and Arctic/North Atlantic Oscillation in the Atlantic; both affect the Caribbean causing multiyear or decadal variation in rainfall and in hurricane occurrence, a dominating factor in the expression of seasonal rainfall patterns.

Rainfall also differs from island to island, coast to coast, lowland to highland, north to south, and east to west. The latitudinal and longitudinal rain gradients may be demonstrated in the Lucayan Archipelago where islands become drier eastward and southward. Grand Bahama Island experiences an average of 1,500 mm of rain; the Turks and Caicos, 600 mm. Within-island variation is often due to wind effects. In much of the Antilles, the easterly trade winds blow fairly steadily at 15–30 km/h bringing ocean-humidified air and rain to the northern and eastern sides of islands. Where mountains block the wind flow, leeward sides are drier. Trade winds pass over low-lying islands. Large islands create their own weather. Dominica has the highest overall rainfall in the Caribbean; on the windward coast, rain totals over 5,000 mm annually; up the mountains, 9,000 mm; but on the leeward west coast, it is 1,800 mm. The pattern is similar on a lesser scale for other mountainous islands. Such differences in annual rainfall within an island go a long way to determine what kinds of plants can grow there and what animals can live there.

Rain has to go somewhere. Throughout the region, but especially in limestone and sandy areas, it may quickly percolate into the ground and find its way to surficial aquifers. The steeper the slope, the less time rain has to be absorbed, and so forms rivers, including spectacular waterfalls, and gentler streams that carry water to lower, flatter areas and to the sea. Rain falling on coastal areas may accumulate into ponds, lakes and reservoirs, or short coastal streams that may flow only seasonally or tidally.

And then there are hurricanes. The hurricane season is more-or-less June through November, but tropical depressions and storms can occur almost any time. Other high rainfall events such as squalls and serious thunderstorms happen more frequently, in some places and some seasons, daily. High rainfall and wind events affect Caribbean vegetation. In 2017, 580 mm of rain fell in one location in Dominica during hurricane Maria. Between 1851 and 2008, 347 hurricanes have passed through the Bahama islands. Different locations have different annual probabilities of experiencing some tropical storm effect. Nassau may be expected to be affected by hurricane force winds on an average of one in 5–6 years and is affected in some way one in every 2–3 years. The Virgin Islands may expect a direct hit 1 in 8 years; Aruba, Bonaire, and Curaçao, one in almost 29 years. Storm-return frequencies are increasing through the Caribbean.

It only takes one Category 3 storm to devastate human-built infrastructure and seemingly devastate an island's vegetation and animal life. Higher categories are even more damaging, as tragically happened in 2017 to Dominica, Puerto Rico, and Saint Croix. Hurricanes can have substantial and immediate effects on plants and animals. Plants may be defoliated, broken up, uprooted, flooded, or killed immediately or sometime later; or their patch of forest may be leveled. Similarly, depending on species and situation, animals may suffer direct mortality from wind, rain, and storm surge and indirect effects from loss of plants, shelter, cover, and food – flowers, seeds, fruits, and invertebrates. Long-term hurricane history affects what native plants and animals might occupy a particular patch of land.

Hurricanes do pose grave risks to small populations of endemic species. Amazon parrots have been repeatedly affected by storms ravaging their montane forests. Recovery of the Dominican amazon (*Amazona imperialis*) (Dominican parrot, imperial parrot, sisserou) and Saint Lucia amazon (*Amazona versicolor*) (jacquot, Saint Lucia parrot) after a 1979 hurricane was a conservation success story. These parrots were again impacted by the 2017 hurricane that devastated their rainforest habitat including in protected areas, Morne Diablotin National Park and the Northern and Central Forest Reserve, requiring active conservation intervention. Puerto Rican parrots (*Amazona vittata*) faced a similar situation when hurricanes in 2017 destroyed tracts of the mountainous rainforest in El Yunque National

Forest. But the species was not at total risk because of captive and reintroduced populations secured elsewhere and artificial feeding. After a hurricane, a banding station at the Ecological Reserve Siboney-Juticí in Santiago de Cuba found that birds were displaced and lost weight following the storm. On the other hand, raptors have an easier time catching prey after a storm clears the vegetation; and some species, such as woodpeckers, benefit from storms in the long-term and can recover their populations within a few years.

Human tragedy, property damage, and economic impact aside, it is worth keeping in mind that hurricanes have been happening since the current climate existed and that the region's native species within their natural habitats have survived these storms. What has changed is that an island's resiliency has declined owing to the reduction and fragmentation of natural vegetation, loss of native hurricane-resistant plant cover, reduced habitat area and population sizes, and wind-prone urban and suburban development, all of which have reduced the capacity of the landscape to support larger populations of plants and animals and to otherwise accommodate hurricane impacts. Nonnative and isolated native trees are toppled, poorly vegetated dunes are swept clean of sand-holding plants, degraded mangrove swamps fail to protect land, and silted up coastal streams inhibit drainage. Recovering nature can help change this.

References

1. Patil, P. G., J. Virdin, S. M. Diez, J. Roberts, and A. Singh. 2016. *Toward a Blue Economy: A Promise for Sustainable Growth in the Caribbean; An Overview*. Washington, DC: World Bank. documents1.worldbank.org/curated/en/965641473449861013/pdf/AUS16344-REVISED-v1-BlueEconomy-FullReport-Oct3.pdf
2. CANARI. 2020. *Lessons Learned from Fisheries-related Livelihoods and Socio-economic Initiatives in the Caribbean*. Barataria, Trinidad and Tobago: Caribbean Natural Resources Institute. canari.org/wp-content/uploads/2019/10/Report_Lessons-from-Carib-fisheries-livelihood-initiativesFinalJul2021.pdf
3. CRFM. 2020. *Caribbean Community Common Fisheries Policy*. Belize: Caribbean Regional Fisheries Mechanism Secretari at. www.crfm.int/index.php?option=com_k2&view=item&id=677:caribbean-community-common-fisheries-policy&Itemid=463
4. United Nations. 2023. *Sustainable Development Goals, Conserve and Sustainably Use the Oceans, Seas, and Marine Resources for Sustainable Development*. New York: United Nations, Department of Economic and Social Affairs. sdgs.un.org/goals/goal14
5. Myers, N., R. Mittermeier, C. Mittermeier, G, da Fonseca, and J. Kent. 2000. "Biodiversity Hotspots for Conservation Priorities." *Nature* 403:853. DOI:10.1038/35002501
6. United Nations Environment Programme. 2023. "Caribbean Environment Programme, Threatened and Endangered Species." www.unep.org/cep/threatened-and-endangered-species
7. Smith, M. L., S. B. Hedges, W. Buck, A. Hemphill, S. Inchaustegui, M. A. Ivie, D. Martina, et al. 2004. "Caribbean Islands." pp. 112–18 in R. A. Mittermeier, P. R. Gil, M. Hoffman, J. Pilgrim, T. Brooks, C. G. Mittermeier, J. Lamoreux, and G. A. B. da Fonseca, eds. *Hotspots Revisited: Earth's Biologically Richest and Most Endangered Terrestrial Ecoregions*. Chicago: University of Chicago Press.
8. Acevedo-Rodríguez, P., and M. T. Strong. 2008. "Floristic Richness and Affinities in the West Indies." *The Botanical Review* 74:5–36.

9. Acevedo-Rodríguez, P., and M. T. Strong. 2012. "Introduction." In *Flora of the West Indies*. Washington, DC: National Museum of Natural History. naturalhistory2.si.edu/botany/WestIndies/
10. Maunder, M., A. Leiva, E. Santiago-Valentín, D. W. Stevenson, P. Acevedo-Rodriguez, A. W. Meerow, et al. 2008. "Plant Conservation in the Caribbean Island Biodiversity Hotspot." *The Botanical Review* 74:197–207.
11. CANARI. 2019. *Ecosystem Profile, The Caribbean Islands Biodiversity Hotspot. Final Version*. Baratara Trinidad and Tobago: Caribbean Natural Resources Institute.
12. González Torres, L. R., A. Palmarola, L. González Oliva, E. R. Becker, E. Testé, and D. Barrios. 2016. "Lista Roja de la Flora de Cuba." *Bissea* 10(Número Especial).
13. Bovarnick, A., F. Alpizar, and C. Schnell. 2010. *The Importance of Biodiversity and Ecosystems in Economic Growth and Equity in Latin America and the Caribbean: An Economic Valuation of Ecosystems*. New York: United Nations Development Programme.
14. CARICOM. 2018. *The State of Biodiversity in the Caribbean Community: A Review of Progress Towards the Aichi Biodiversity Targets*. Georgetown, Grenada: Caribbean Community Secretariat.
15. Raffaele, H. A., J. Wiley, A. Keith, and J. I. Raffaele. 2020. *Birds of the West Indies*, 2nd edition. Princeton: Princeton University Press.
16. Faaborg, J. 2002. *Saving Migrant Birds, Developing Strategies for the Future*. Austin: University of Texas Press.
17. La Sorte, F. A., D. Fink, W. M. Hochachka, and S. Kelling. 2016. "Convergence of Broad-scale Migration Strategies in Terrestrial Birds." *Proceedings of the Royal Society B* 283:20152588. DOI:10.1098/rspb.2015.2588
18. Fritts, R. 2022. "Avian Superhighways: The Four Flyways of North America." American Bird Conservancy. abcbirds.org/blog/north-american-bird-flyways/#:~:text=Birds%20navigate%20along%20more%20or,%2C%20Central%2C%20and%20Pacific%20Flyways
19. CAF. 2022. "New Initiative to Protect Bird Migration Flows in the Americas." CAF Development Bank of Latin America. www.caf.com/en/currently/news/2022/12/new-initiative-to-protect-bird-migration-flows-in-the-americas/?parent=6571
20. Bramwell, D., and J. Caujapé-Castells, eds. 2011. *The Biology of Island Floras*. Cambridge: Cambridge University Press.
21. Santiago–Valentin, E., and R. G. Olmstead. 2004. "Historical Biogeography of Caribbean Plants: Introduction to Current Knowledge and Possibilities From a Phylogenetic Perspective." *Taxon* 53:299–319. DOI:10.2307/4135610
22. Nieto-Blázquez, M., A. Antonelli, and J. Roncal. 2017. "Biogeography of Endemic Seed Plant Genera in the Caribbean: Did GAARlandia Play a Role?" *Ecology and Evolution* 7:10158–74. DOI:10.1002/ece3.3521
23. Stattersfield, A. J., M. J. Crosby, A. J. Long, and D. C. Wege. 1998. *Endemic Bird Areas of the World: Priorities for Biodiversity Conservation*. Cambridge: BirdLife International.
24. Presley, S. J., and M. R. Willig. 2008. "Composition and Structure of Caribbean Bat (Chiroptera) Assemblages: Effects of Inter-island Distance, Area, Elevation, and Hurricane-Induced Disturbance." *Global Ecology and Biogeography* 17:747–57.
25. Wege, D. C., and V. Anadon-Irizarry. 2008. *Important Bird Areas in the Caribbean: Key Sites for Conservation*. Cambridge: BirdLife International.
26. Munch, P., P. O. Antoine, and B. Marcaillou. 2020. "By Land or Sea: How Did Mammals Get to the Caribbean Islands?" EOS 101. eos.org/science-updates/by-land-or-sea-how-did-mammals-get-to-the-caribbean-islands
27. Iturralde-Vinent, M. A., and R. MacPhee. 1999. "Paleogeography of the Caribbean Region: Implications for Cenozoic Biogeography." *Bulletin of the American Museum of Natural History* 238:1–95.

28. Cooke, S. B., L. M. Davalos, A. M. Mychajliw, S. T. Turvey, and N. S. Upham. 2017. "Anthropogenic Extinction Dominates Holocene Declines of West Indian Mammals." *Annual Review of Ecology, Evolution, and Systematics* 48:301–27. DOI:10.1146/annurev-ecolsys-110316-022754
29. Waites, D., and I. Rouse. 1989. "Environmental Diversity and Maritime Adaptations in the Caribbean Area." pp. 129–44 in P. E. Siegel, ed. *Early Ceramic Population Lifeways and Adaptive Strategies in the Caribbean*. Oxford: BAR.
30. Rouse, I. 1992. *The Taínos. Rise and Decline of the People Who Greeted Columbus*. New Haven: Yale University Press.
31. Fernandes, D. M., K. A. Sirak, H. Ringbauer, J. Sedig, N. Rohland, O. Cheronet, and M. Mah. 2021. "A Genetic History of the Pre-contact Caribbean." *Nature* 590:103–10. DOI:10.1038/s41586-020-03053-2
32. Steadman, D. W., and J. Franklin. 2020. "Bird Populations and Species Lost to Late Quaternary Environmental Change and Human Impact in the Bahamas." *Proceedings of the National Academy of Sciences* 202013368. DOI:10.1073/pnas.2013368117
33. McGinley, K. 2023. "The Caribbean." USDA Forest Service, International Institute for Tropical Forestry. sites.google.com/site/forestryencyclopedia/Home/The%20Caribbean
34. Keegan, W. F., and L. A. Carlson. 2008. *Talking Taino: Caribbean Natural History From a Native Perspective*. Tuscaloosa: University of Alabama Press.
35. Fitzpatrick, S. M. 2013. "Seafaring Capabilities in the Pre-Columbian Caribbean." *Journal of Maritime Archaeology* 8:101–38.
36. Turvey, S. T. 2009. *Holocene Extinctions*. Oxford: Oxford University Press.
37. Walters, B. B. 2019. *The Greening of Saint Lucia: Economic Development and Environmental Change in the West Indies*. Kingston: University of West Indies Press.
38. Taylor, M. A., T. S. Stephenson, S. Tannecia, A. A. Chen, K. A. Stephenson. 2012. "Climate Change and the Caribbean: Review and Response." *Caribbean Studies* 40:169–200. www.redalyc.org/articulo.oa?id=39226915007
39. Moore, W., W. Elliot, and T. Lorde. 2017. "Climate Change, Atlantic Storm Activity and the Regional Socio-economic Impacts on the Caribbean. *Environment, Development and Sustainability* 19:707–26. DOI:10.1007/s10668-016-9763-1
40. Reguero, B. G., I. J. Losada, P. Díaz-Simal, F. J. Méndez, and M. W. Beck. 2015. "Effects of Climate Change on Exposure to Coastal Flooding in Latin America and the Caribbean." *PLOS ONE* 10(7):e0133409.
41. Beckford, C. L., and K. Rhiney. 2016. "Geographies of Globalization, Climate Change and Food and Agriculture in the Caribbean." In C. L. Beckford and K. Rhiney, eds. *Globalization, Agriculture and Food in the Caribbean*. London: Palgrave Macmillan. DOI:10.1057/978-1-137-53837-6.
42. PRCCC. 2022. *Puerto Rico's State of the Climate of 2014–2021*. San Juan: Puerto Rico Climate Change Council. www.pr-ccc.org/wp-content/uploads/2023/06/PR_StateOfTheClimate_2014-2021_PRCCC-09-2022-1.pdf
43. Areces-Mallea, A. E., A. S. Weakley, X. Li, R. G. Sayre, J. D. Parrish, C. V. Tipton, and T. Boucher. 1999. "A Guide to Caribbean Vegetation Types: Preliminary Classification System and Descriptions." The Nature Conservancy. www.conservationgateway.org/ConservationByGeography/NorthAmerica/Documents/A-Guide-to-Caribbean-Vegetation-Types.pdf
44. Barbour, W. R. 1942. "Forest Types of Tropical America." *Caribbean Forester* 3:137–50.
45. Beard, J. S. 1949. *The Natural Vegetation of the Windward and Leeward Islands*. Forestry Memoirs 21. Oxford: Clarendon Press.
46. Beard, J. S. 1955. "The Classification of Tropical American Vegetation Types." *Ecology* 36:89–100.

47. Holdridge, L. R. 1967. *Life Zone Ecology*. San Jose: Tropical Science Center.
48. McGinley, K. 2023. "The Caribbean." In F. Cubbage, ed. *Forest and Forestry in the Americas: An Encyclopedia*. Society of American Foresters and International Society of Tropical Foresters. sites.google.com/site/forestryencyclopedia/Home
49. Ahmad, N. 2011. *Soils of the Caribbean*. Kingston: Ian Randle Publishers.
50. Muhs, D. R., C. A. Bush, K. C. Stewart, T. R. Rowland, and R. C. Crittenden. 2017. "Geochemical Evidence of Saharan Dust Parent Material for Soils Developed on Quaternary Limestones of Caribbean and Western Atlantic Islands." *Quaternary Research* 33:157–77.
51. World Meteorological Organization. 2023. "State of the Climate in Latin America and the Caribbean 2022." Geneva: WMO. library.wmo.int/index.php?lvl=notice_display&id=22309
52. Osborn, L. 2023. "Caribbean Weather: Annual Temperature & Rainfall." www.currentresults.com/Weather/Caribbean/average-annual-temperature-rainfall.php
53. Taylor, M. A., and E. J. Alfaro. 2005. "Central America and the Caribbean." In J. E. Oliver, ed. *Encyclopedia of World Climatology*. Dordrecht: Springer. DOI:10.1007/1-4020-3266-8_37

2 Recovering nature

The Caribbean environment today is much different than when Indigenous Peoples began arriving from South America. With environmental impacts increasing exponentially during colonial and modern eras, the alteration of the Caribbean's terrestrial environment, protected areas aside, is to a large extent nearly complete. The nature of the apparently green Caribbean we see today, however diverse, endemic, and globally significant it may be, is but a shadow of the Caribbean that existed before humans arrived. But even so, experience has shown that Caribbean lands released from intensive use have impressive regenerative capacities. Much of Caribbean nature can be purposefully recovered.

2.1 Nature recovery in principle and practice

Nature left alone will affect some degree of recovery from disturbance on its own. An empty lot or pasture will be taken over by weeds, then by shrubs and trees of one sort or another; a hurricane-downed patch of forest will undergo a natural succession starting with pioneering plant species adapted to claim newly opened space. But given an island's limited landscape, recovering nature usually needs to be a deliberate process controlled by owners, land managers, or restoration practitioners. Recovering indigenous nature, or ecological restoration, in its most precise sense involves purposeful intervention to assist the restoration of an ecosystem that has been degraded, damaged, or destroyed.[1] Recovering nature fundamentally involves employing ecological principles to assemble communities of native plants and the animals they support. Restoration may be in response to natural forces such as fires, floods, volcanoes, or hurricanes. Or, restoration may be in response to a range of human-caused alterations from homes and farms on small private holdings to more extensive developments such as large-scale agriculture, pastures, tree plantations, and residential, commercial, and infrastructure development.

International goals encourage nations to protect, restore, and promote sustainable use of terrestrial ecosystems, sustainably manage forests, combat desertification, reverse land degradation, and halt biodiversity loss.[2] Such globally articulated goals resonate in the Caribbean where the region houses significant internationally recognized protected areas and where environmental sustainability is accepted as crucial to the region's economic base. Most Caribbean nations have agreed to these goals. Achieving sustainable development usually requires active intervention to protect and recover parts of the natural environment and the ecosystem services they provide.

DOI: 10.1201/9781003396321-3

Ruins of a plantation-era waterwheel being overtaken by recovering forest on Tobago.

Such intervention to recover nature has many nuanced names including recovery, restoration, re-creation, rehabilitation, remediation, rewilding, and regreening – covering a range of restoration activities. The UN Decade on Restoration has identified this range as including the reduction of negative environmental and societal impacts such as pollution and unsustainable resource use and management; remediation of contaminants, pollutants, and other threats; rehabilitation of ecosystem functions and services in highly modified areas such as former mining sites and degraded production systems; and ecological restoration that removes degradation and assists in recovering an ecosystem to the trajectory it would be on if degradation had not happened.[1, 3] Regreening tends to refer to widespread planting of trees, shrubs, and grasses and other site improvements on degraded habitat. Rewilding tends to refer to establishing naturalistic habitat. In this book, the terms recovery and restoration are used interchangeably and incorporate concepts such as rehabilitation, ecosystem restoration, recovery, re-creation, and remediation.

A restoration might be implemented at any scale from a large forest reserve, to small patches of habitat such as a public park, botanic garden, town lots, and family farm plots, to even smaller areas such as private yards and gardens. Restoration, particularly ecosystem restoration, tends to refer to actions to recover or re-create large-scale ecosystems. Restorative action on smaller plots, especially in towns and cities, is often referred to as regreening, habitat enhancement, or natural gardening. By whatever term and at whatever scale, recovering nature aims to create ecosystems, habitats, and plant and animal communities that allow for a more-natural trajectory of ecological change.

A mangrove restoration project on Grenada.

Ecosystem is a useful concept in recovering nature, one well understood having been taught in schools for decades. An ecosystem may be defined at any scale from a water droplet, to a mountain-covering tropical forest, to an entire island. At whatever scale, an ecosystem consists of its biodiversity (such as plants, animals, fungi, and bacteria), its physical environment (such as water, soils, geology, chemicals, and climate), and processes by which energy and materials flow within, into, and out of its specified boundaries. Upon recovery, an ecosystem would exhibit more characteristics of the natural target ecosystem including biodiversity, ecosystem functions, sustainability, and resilience. However, only in the largest and most intact locations can an approach to this result be expected. On smaller sites, only partial recovery of an ecosystem and its ecological services can be expected.

Crucial to recovering nature is defining a reference, a target ecosystem that illustrates goals for restoration and also takes into account future environmental change.[1] Defining a reference condition might include information from a similar intact ecosystem, a description of past ecosystems that could be expected to redevelop, or descriptions or models of intended endpoints. In the Caribbean, establishing an ecosystem's undegraded condition may not be fully possible owing to centuries of environmental change. In general, establishing a target reference condition would involve a combination of estimating an undegraded condition, understanding comparable ecosystems, appreciating an array of ecological and historical information about the land, incorporating current aspirations for the restoration, and estimating future environmental change. Once a target reference condition is established, specific project goals and measurable objectives can be established.

Tropical montane forest of the Blue and John Crow Mountains National Park in Jamaica, a World Heritage Site and biodiversity hotspot, an example that might serve as a reference for forest restoration.

As commonly understood, the words *restoration* or *recovering* might infer bringing something back to a like-new condition, such as restoring a damaged antique table. Restoration or recovering the environment is something different; there is no going back completely. Restoration's goal is to assist the recovery of nature to a condition that allows the system to evolve and adapt with minimal human management. Recovering nature creates a natural trajectory rather than an end product. The goal of full recovery is not often possible, especially on the small land masses of the Caribbean because of the need to continue mitigating past and future anthropogenic stresses. Most sites under restoration, as a practical matter, continue to require some level of active management requiring money, people, and long-term commitment.

Recovering nature is not so much a thing as a process, one that involves planning, consultation, implementation, assessment, and adaptation. For large projects, the process would encourage participation by stakeholders including local residents and interest groups. Its administrative and operational processes are long-term, likely indefinite. At the opposite end of the spectrum, for a small garden, the process might involve just thinking through the goals in advance. Since recovering nature does not result in a static situation, the process involves allowing for ongoing change and evolution. A volcano or mountain landslide may start the recovery process from bare ground or a hurricane may alter conditions such that an entirely different ecosystem follows. Even for the largest forests, future conditions will change the recovery trajectory. At the other end of the scale, gardens too are never static, despite how traditional gardening has tried to make it so.

Active management of nature recovery can take many forms, and many guides and models are available.[1, 3–9] A core feature of any restoration is encouraging, or if necessary, re-establishing native vegetation appropriate to the site. Restoration projects may be extensive, such as creating a treed landscape to reverse desertification in Africa or mitigating anticipated climate change by restoring coastal mangroves and marshes in Cuba, Granada, and elsewhere.[10] Recovery of large areas of forest from hurricanes may involve some replanting, but most of the recovery is often best left to nature, if only because of its extensiveness. Active restoration can be particularly effective at smaller scales such as restoring coastal dunes, offshore islands after invasive mammal removal, and re-creating native plant communities in otherwise highly landscaped sites, such as a public park. It may be replacing nonnative trees and other plants with native species in a farm, yard, or garden setting. Beyond plants and associated organisms, restoration focuses on removing impediments to natural recovery trajectories such as eliminating incompatible land or resource use, changing forestry practices, rerouting roads, removing dams, reducing nonnative plants and animals, introducing or suppressing fire, establishing more-natural hydrological management, reintroducing native wildlife, establishing monitoring and adaptive management systems, encouraging local co-management, and finding secure funding – whatever is needed to restore ecological processes and native biotic communities.

Because forests once dominated the Caribbean terrestrial landscape, restoration of forest communities is a high priority. Caribbean forestry practices benefit from having

Holistically managed rainforest in El Yunque National Forest, Puerto Rico.

an extensive long-term scientific basis and engagement with the international forestry commuity.[11–14] Commercial forestry is decreasing as a percentage of national economies with fewer than half of Caribbean islands having active timber industries, including Martinique, Cuba, Haiti, and The Bahamas.[11] Nonetheless, almost every country has reforestation, afforestation, or tree plantation projects ongoing. Although forests traditionally were managed primarily for wood production, they also may provide recreation, hunting, gathering, grazing, and imbedded small-plot agriculture. Increasingly, environmental values such as soil and water conservation, biodiversity conservation, animal habitat management, carbon sequestration, and cultural heritage preservation are being appreciated. These expanded considerations set the stage for managing forests holistically and sustainably for their multiple ecosystem services.

Deforestation does continue, depending on country, under such pressures as residential development, small-plot agriculture, and fuelwood and charcoal harvest. Although Caribbean nations have forest-management policies and laws, funding, technical services, and personnel are usually insufficient. Nonetheless, throughout the region, governments and local stakeholders are increasingly focused on forest resilience, their breadth of values and services, and roles in sustainable development.

Of the 10% of the Caribbean that is land, only 10% still supports a semblance of native plant and animal communities.[14] Most is in large protected areas, which provide the greatest opportunities for recovering nature and are key to biodiversity conservation. But there are limits to creating large reserves. Demand for housing, commerce, and tourism leaves few opportunities for new large-scale nature restoration in most nations, especially on small islands. But, recovering Caribbean nature need not only be about

Coastal scrub vegetation, an example of the limited areas still supporting native plant communities.

A guesthouse nestled within thatch palm scrub forest.

restoring a large rain or scrub forest; it is also about protecting small reserved areas of high biodiversity, setting aside portions of public parks and botanical gardens as natural areas, regreening town landscapes, planting highway margins, or creating naturalistic private gardens with indigenous plants. Recovering Caribbean nature creates a matrix of more-or-less naturalistic habitats of various sizes scattered about.

The terrestrial Caribbean is thoroughly a culturally derived landscape; even its impressive forests are almost entirely secondary growth. So, in restoring nature, ongoing cultural and economic influences must be taken into consideration. Human activity created the current terrestrial environment, and any recovered habitat will continue to experience human engagement. In many situations, this continued human use can be done in sustainable ways. Forests often have remnant previously planted nonnative trees, the use of which in traditional ways may make conserving the forest more acceptable to the nearby community without compromising the forests' trajectory toward recovering a more-natural ecosystem. Taking into consideration local and community interests and expectations and providing opportunities for engagement have proven to be essential to accomplish restoration. Traditional knowledge about using natural resources sustainably can, in fact, help guide restoration. Bush medicine, for example, is a historically important cultural tradition usually involving the use of native plant species. Conservation of heritage plants can be an exceptionally useful component of nature restoration in the Caribbean.

As usually designed, commercial and public works require continuing high maintenance; yet, there is never enough money, peoplepower, or political will to keep such maintenance going in the long run. Almost always, the original creative design eventually gets lost on the ground and in collective memories, allowing nonnative weeds and abusive uses to take over. A naturalistic design using native plants imitating native habitats, on the other hand, is mostly self-sustaining and less expensive to manage indefinitely. Tourist attractions landscaped in environmentally sensitive ways can be publicized as showcasing an island's unique natural heritage as well as costing less to plant and maintain, saving irrigation water, and better enduring storms. A matrix of naturalistic plots incorporating public and private land within a sea of other land uses is likely the ideal solution for recovering nature in the Caribbean.

2.2 Attributes of recovering nature

How to know when recovery is done? A positive comparison with the reference ecosystem is the ultimate test; but that takes a long time. And that endpoint may or may not be reached other than in special circumstances. However, there are attributes of recovering nature that can be examined along the way, and they can guide the recovery process.[15]

2.2.1 Naturalistic

To the extent possible, a restoration site, park, or natural garden re-creates natural habitat through its plant species, vegetation structure, animal species supported, and through interactions among its plants, animals, microbes, and the physical environment. Protection, management, and restoration of large plots such as parks, forests, and other protected areas can attempt to achieve an environmental trajectory that matches its target reference. Goals for recovering smaller patches need to be more modest. But at both ends of the spectrum, the goal is to create a situation that better reflects the natural habitat that birds and other native animal species can use.

Usual land uses such as urban and suburban development seldom imitate nature. Most tourist infrastructure, intense agricultural fields, improved pastures, monoculture lawns, agroforest plantations, and formal gardens are far from being natural. More-natural environments are neither monocultures nor overly designed but rather are plots of mixed species, crowded, competing, interlocking, dense, and mostly indigenous. They have forest trees of height and species appropriate to the location. They have shade-tolerant plants and ferns below the canopy and sun-loving plants around the outside. They have vines tangling through the trees, air plants and orchids affixed to the trunks. They have a self-produced ground substrate of leaves, sticks, and other natural organic matter. They are on soils and have a water budget appropriate to the location. They are not managed for the pampering of individual plants as individual plants may come and go as the plot follows its trajectory. Management is concerned with the ecosystem as a whole, with the plants themselves eventually deciding the character of the habitat. The creation of a naturalistic habitat rather than the planting and caring for individual plants is an essential principle of recovering nature.

2.2.2 Ecological

A restoration must be designed in accord with principles of ecology to create a more naturally functioning ecosystem. Natural communities grow, change, and experience succession as species replace others over time. Natural communities consist of collections of interacting species bound together by ecological processes such as predation, herbivory, competition, commensalism, and energy flows. Predation, for example, is a natural process, and natural areas provide insects, spiders, snails, seeds, and fruits as food for birds and other animals. They also provide small birds to be eaten by hawks and falcons. In the tropics, plants have differing cycles of reproduction, some flower all year, some in dry season, some in wet, together providing a succession of food for birds and other wildlife throughout the entire year. Animals compete with each other for resources and space; plants compete with each other for light, nutrients, and space. Some organisms cooperate synergistically, such as nitrogen-fixing bacteria and soil fungi that make nutrients available to plant roots or ants guarding their host tree. Plant communities in restorations evolve, and succession may be periodically reset by disturbances such as floods, droughts, fires, hurricanes, and human activities. Native plants in gardens will similarly change over time in the plants and in the animals they attract. Like nature, natural areas are not static but accommodate to and even thrive on natural change.

2.2.3 Diversity

Diversity in part means the richness of species to be found in a restoration site, park, or natural garden. It also means genetic diversity of the indigenous populations of organisms naturally occurring there. As different plants make shelter and food available at different times and seasons, sometimes in different years, the more plant diversity, the more diverse the food and shelter provided for birds and other animals and the more complex and resilient will be the community as a whole. Diversity also includes physiognomy of the patch, its structure, and microhabitats. Trees, shrubs, vines, ground cover, natural mulch, grasses, and water sources all contribute to structural and functional diversity. Diversity is both horizontal (covering the ground) and vertical (layering above ground). Various plants make up tiers within the canopy, creating vertical diversity. The more diverse the species, structure, design, and features, the better they are for approaching a complete, resilient community.

2.2.4 Native

Recovered nature is native. Caribbean islands are each blessed with an extensive native flora – the list of plants indigenous to each island. Native plants provide a diversity of food, shelter, and structure in ways that nonnative plants seldom can. Native forest plants naturally grow clustered together, including massive trees with big canopies and some with buttressed trunks, understory trees, epiphytes, shrubs, vines, and herbaceous plants, all of which provide structure and living space for animals. Even in xeric habitats where plants space themselves, they interact if only at their roots. Native plants flower and fruit in concert with the annual cycles of native animal species. They host species and communities of native insects, spiders, snails, and other invertebrates, and native birds, lizards, frogs, and bats. To accomplish nature recovery, plants used in restoration reflect the native plant communities of the area.

Native plants are adapted to their native environment. Once established, they generally require little or minimal care and little or no expense. Adapted to local seasonality, they need no artificial watering. Adapted to the local soils, they need no fertilizer.

A spectacled thrush with a bill full of worms collected within a natural garden setting in Tobago.

Adapted to withstand chewing and probing of local insects, they need no pesticides. In a garden, what non-xeric tropical native plants need most often is pruning to keep their relative exuberance under control. In contrast, most noninvasive nonnative plants are expensive to maintain. Bred for horticultural purposes not for local conditions, they may require chemical pest control, soil modification, irrigation, and fertilizers. They do poorly in tight groupings, and so, to avoid competition and allow horticultural care, they are often arranged spread out as specimen trees and shrubs where they do not contribute to creating nature-emulating plant communities. And specimen plants get blown down in hurricanes more readily than native plants interwoven in native communities.

Since a native species or its ancestors may have arrived thousands or tens of thousands of years ago, time has allowed for genetic diversification. A population being indigenous to an individual island may be genetically different from the same species on other islands. The more local a particular specimen, the more likely that it represents the local gene pool – and possesses local adaptations. There is an unfortunate tendency to think that any plant or animal one grew up with, found in the wild or prominently, is native; this is not so. Although perhaps existing on the islands for centuries and introduced for good reasons, most nonnative plants have limited ecological value. Nonnatives take space that a native species could occupy while providing their ecological benefits. In the context of recovering nature, nonnative plants tend not to provide the food that native birds and other animals need, as they did not coevolve.

That said, nonnative plants are not all wrong in all cases at all times. They are certainly not inappropriate in agriculture, vegetable gardens, and fruit orchards. Species related to native species may serve as host plants for native butterfly and moth caterpillars and

Landscaping that reflects the native plant communities of the area.

other species. Plants cultivated for human food tend to have accumulated lots of insect "pests," which can be food for native animals. Some nonnative flowers are accessible to native nectar-feeders. A few traditionally used nonnative plants, such as sapodilla, avocado, mango, or cacao, imbedded within a forest, may not seriously disrupt ecosystem function yet contribute to community welfare. Human fruit can also be bird fruit. But the list of accommodating nonnative species is slim compared to the diversity of the natural Caribbean flora.

Of all the native animals recovering nature might support, among the more obvious and appreciated are birds. Large-scale restoration can re-create or maintain communities of birds, and it is here where indigenous and endemic species might best be conserved. Common birds and many winter residents are some of the main beneficiaries of recovering smaller patches of habitat such as scrub, small public parks, gardens, and commercial settings. At the small-plot and garden scale, providing habitat for birds might be a primary goal for creating a more-natural environment and a measure of its success.

2.2.5 Connectivity

Connecting restorations through corridors or other means is a key to island-wide nature recovery. Every patch of restored or regreened land is valuable but more so to the extent that animals and plant propagules can move among them. The matrix formed by natural habitat from forest reserves to backyard gardens expands the footprint of recovered nature that supports a species or population. Between patches, likely, are various degrees of inhospitable landscape. Birds and bats may be able to navigate between distant patches

Some culturally traditional food plants, such as this papaya, can be imbedded into the forest without disrupting the plant community.

as may butterflies and some other insects. Frogs, lizards, turtles, and other terrestrial animals may have more difficulty. Connections among natural areas are called corridors; they allow recovered patches to function more broadly across an island's landscape.

References

1. Gann, G. D., T. McDonald, B. Walder, J. Aronson, C. R. Nelson, J. Jonson, J. G. Hallett, et al. 2019. "International Principles and Standards for the Practice of Ecological Restoration. Second Edition." *Restoration Ecology* S1–S46. DOI:10.1111/rec.13035
2. United Nations. date unknown. *Sustainable Development Goals, 15*. New York: United Nations, Department of Economic and Social Affairs. sdgs.un.org/goals/goal15
3. FAO, IUCN, CEM, & SER. 2021. *Principles for Ecosystem Restoration to Guide the United Nations Decade 2021–2030*. Rome: FAO. www.fao.org/3/cb6591en/cb6591en.pdf
4. Perrow, M. R., and A. J. Davy, eds. 1999. *Handbook of Ecological Restoration. Volume 1. Principles of Restoration*. Cambridge: Cambridge University Press.
5. Ramsar Convention on Wetlands. 2002. "Principles and Guidelines for Wetland Restoration." www.ramsar.org/key_guide_restoration_e.htm

6. Higgs, E. 2003. *Nature by Design: People, Natural Processes, and Ecological Restoration*. Cambridge: MIT Press.
7. Canadian Parks Council. 2008. "Principles and Guidelines for Ecological Restoration in Canada's Protected Natural Areas." Parks Canada and the Canadian Parks Council. parks.canada.ca/nature/science/conservation/ie-ei/re-er/pag-pel
8. Cairns, S., N. Dudley, C. Hall, K. Keeneleyside, and S. Stolton. 2012. "Ecological Restoration for Protected Areas: Principles, Guidelines and Best Practices." IUCN. www.iucn.org/resources/publication/ecological-restoration-protected-areas-principles-guidelines-and-best
9. Janishevski, L., G. Santamaria, S. Gidda, and H. D. Cooper. 2015. "Ecosystem Restoration, Protected Areas and Biodiversity Conservation." *Unasylva* 245:19–28.
10. UNDP Climate Change Adaptation. 2023. "Coastal Resilience to Climate Change in Cuba through Ecosystem Based Adaptation ('MI COSTA')." www.adaptation-undp.org/GCF-Cuba
11. FAO. 2023. "Latin American and Caribbean Forestry Commission (LACFC). " www.fao.org/americas/coflac/en/
12. UN Food and Agricultural Organization. 2020. "Global Forest Resources Assessment 2020. Main Report." DOI:10.4060/ca9825en.
13. McGinley, K. 2008. "The Caribbean." In F. Cubbage, ed. *Forests and Forestry in the Americas: An Encyclopedia*. Society of American Foresters and International Society of Tropical Foresters. sites.google.com/site/forestryencyclopedia/Home/The%20Caribbean
14. CEPF. 2010. *Ecosystem Profile: The Caribbean Islands Biodiversity Hotspot*. Arlington: Critical Ecosystem Partnership Fund.
15. Kushlan, J. A., and K. Hines. 2014. *Attracting Birds to South Florida Gardens*. Gainesville: University Press of Florida.

3 Place-based conservation

Conservation is ultimately place-based, protecting and managing sites where native plants and animals live. The goal is to secure native populations in their appropriate habitats – that is, in "places." Depending on the species and situation, other conservation measures might also be needed such as appropriately regulating killing, the pet trade, harassment, and exports and imports, providing nesting sites and supplemental feeding after storms, or maintaining ex situ conservation populations. Although international, national, and private conservation programs provide guidance, frameworks, and encouragement, it is conservation on the ground that preserves biodiversity. Conservation actions for "places" are local. Recovering nature is accomplished place by place, park by park, garden by garden, and a bit of regreening by a bit of regreening. Or, as expressed in the Introduction, creating a patch-by-patch approach to recovering Caribbean nature.

3.1 Protected areas

International programs explicitly recognize specific places for their biodiversity or ecological values and in doing so highlight their special priority for place-based conservation action. Mentioned previously are the Caribbean's hundreds of Key Biodiversity Areas, Important Bird Areas, and Endemic Bird Areas.[1–7] Alliance for Zero Extinction sites recognize places supporting the world's most endangered species; over 30 have been identified in the Caribbean.[8, 9] The Caribbean's over 40 Biosphere Reserves recognize protected areas demonstrating sustainable relationships between people and nature.[10] Five of the Caribbean's World Heritage Sites are recognized for their natural heritage: Pitons Management Area, Saint Lucia; Morne Trois Pitons National Park, Dominica; Blue and John Crow Mountains, Jamaica; and Alejandro de Humboldt National Park and Desembarco del Granma National Park, Cuba.[11] Nineteen Caribbean sites are recognized as wetlands of international importance by the Convention on Wetlands; most have terrestrial components.[12] So, there is a catalogue of hundreds of places identified as key for conserving terrestrial biological diversity in the Caribbean.

Each Caribbean country and territory has set aside protected areas and institutionalized their management.[13–16] Some of the internationally recognized places may be officially established as protected areas, but not all enjoy such recognition. However, officially protected areas in the Caribbean number over 500; and the list is growing.[16] Such protected sites have many names. The most famous are national parks, which tend to incorporate

DOI: 10.1201/9781003396321-4

Exuma Cays Land and Sea Park, a protected area in The Bahamas, established in 1958 as the first marine reserve in the Caribbean, also incorporating terrestrial habitats on the included cays.

Blue and John Crow Mountains National Park and Forest Reserve protected area in Jamaica.

both conservation and visitation. Others (in their English versions) may be a biological reserve, forest reserve, integral biological reserve, national forest, natural monument, natural resources protected area, national nature reserve, open space, plant preserve, private reserve, protected area, protected forest, protected watershed, regional nature park, reserve, scientific reserve, wildlife refuge, all of which are used in the Caribbean. There also is a global nomenclature for protected areas based on their management (which may or may not be the formal name) – strict nature reserve, wilderness area, national park, natural monument or feature, habitat or species management area, protected landscape or seascape, and protected area with sustainable use of natural resources.[17] The effectiveness of these protected areas does differ as some reserves enjoy substantial support, some are quasi-parks with much human use, and others exist principally on paper.[18]

Terrestrial protected areas have a long and distinguished history in the Caribbean, dating back two and a half centuries. The Main Ridge Forest Reserve on Tobago (1776) and King's Hill Forest Reserve on Saint Vincent (1791) are among the oldest such reserves. Cuba's first nature reserve was the Botanic Garden of Havana in 1817, followed by Cayo Mal País and Cayo Buenavista in the Gulf of Batabanó in 1923, and Sierra Cristal National Park in 1930. In Puerto Rico, El Yunque National Forest/Luquillo Experimental Forest got its start in 1876 when the Spanish crown set aside a forest reserve in the Luquillo Mountains, a reservation confirmed in 1903 when the United States established the Luquillo Forest Reserve. Among early wildlife reserves is Culebra National Wildlife Refuge in Puerto Rico (1909). Grand Etang Forest Reserve on Grenada, now attached to the Annandale Forest Reserve, was established in 1910.

Some Caribbean countries and territories already have met the 2030 goal of having 30% of their land in protected status, including The Bahamas, Trinidad and Tobago, and Turks and Caicos. Cuba protects over 200 terrestrial sites. In contrast, others such as Barbados, the British Virgin Islands, Grenada, Haiti, Puerto Rico, and Sint Maarten have so far been able to protect less than 10%.[15] Of remaining forest area in the Caribbean, as of 2015, 59% was under protected status.[19] Although the proportion of protected forests remains low in parts of the Greater Antilles (Puerto Rico, Haiti), it is quite high on some of the Lesser Antilles islands (Martinique, Guadeloupe, Dominica).[20] On highly developed islands, protecting 30% of the land may be unrealistic; but overall, the Caribbean has an exceptional inventory of protected areas, providing the place-based geographic backbone for conserving and recovering Caribbean nature.

There is a substantial literature of technical information on identifying, planning, managing, and evaluating parks, reserves, and forests and on sustainability from global, regional, and national perspectives.[21–24] And, there is similarly substantial international engagement in protected area restoration in support of global biodiversity and restoration goals.[13–15, 25–29] One of the early 30 × 30 flagship programs is the Small Island Developing State restoration program that includes Saint Lucia.[30] Most Caribbean nations have national management planning for their declared reserves and forests, and many sites have local management plans.

Owing to the Caribbean's history of habitat degradation, few, if any, protected areas can be expected to have a complete, unaltered environment and many face ongoing degradation

Louvet Estate on Saint Lucia protects endemic iguanas and their beach, managed as a partnership between the property owners and the Saint Lucia Forestry Department.

through incompatible uses, such as farming and pasturage. So, nearly all terrestrial protected areas require some level of ecosystem restoration to achieve greater ecological integrity. Depending on the nation or territory and the history, management of various protected areas may be expected to differ. Management might be assigned to one or more government agencies or nongovernmental organizations, be comanaged through shared governance, be privately held, or be managed by local communities, including indigenous communities. The leadership of local non-governmental institutions can be the key to long-term success.

Depending on the country, nongovernmental organizations participate in various ways. Government-sharing governance over protected areas, especially with local stakeholders, is considered internationally to provide more equitable, effective, efficient, and fair management of conservation sites than government management alone.[14] The Bahamas National Trust and Jamaica Conservation and Development Trust are examples of nongovernmental organizations that manage national parks in their countries at the behest of and in cooperation with national government agencies. Additionally, in recent years, it has come to be accepted that there are ways other than government-run reserves to conserve terrestrial habitat, called Other Effective Area-based Conservation Measures or OECMs. The El Zorzal Private Reserve in the Dominican Republic is an example of a private reserve. The list and functioning of protected areas and OECMs are monitored at the global scale by multinational organizations so that protected area management can be assessed against expectations and goals.[14]

Global standards are that successful conservation is to be accomplished through a combination of governance, planning, and management.[2] Whatever governance, oversight mechanisms, or local management is employed at a particular location, the goal would be maintaining and recovering nature to the extent possible while managing sustainably for human and ecological benefits, within the constraints imposed on each protected area. Nearly all Caribbean nations and territories have agreed to the Global Goals for Sustainable Development. In some countries, such international agreements have become law. Sustainable development requires sustainable nature; so, achieving sustainability is a key concept for conserving nature on land. Such international standards and structure provide multiple pathways for support and justification for conserving and recovering protected areas and other efforts to regreen the Caribbean landscape.

3.2 Public and botanic gardens

An important subset of the Caribbean's protected places is its public and botanic gardens. Horticultural gardens, public and private, are an integral part of Caribbean land-use history.[31–33] Most countries have botanic gardens; Cuba has 13. The oldest, the Saint Vincent and the Grenadines Botanic Garden, dates from 1765. Whether run by governments, organizations, or privately, such gardens have historically played and continue to play important societal and scientific roles. Historically, colonial governments, private collectors, and European botanic gardens established gardens and plant introduction stations reciprocally in their colonies and Europe. Individual collectors and botanic gardens in the colonies supplied native Caribbean plants to the hothouses of European botanic gardens and private collectors. Meanwhile, temperate plants and tropical plants collected elsewhere were sent to Caribbean gardens for introduction and transfer among colonies. By the late 1700s, Joseph Banks, the Royal Botanic Garden at Kew, and Chelsea Physic Garden were predominant in influencing plant-species dispersion around the tropics, including through the Caribbean.[33, 34] With the decline of the plantation economy, Caribbean

Nassau Botanical Gardens, The Bahamas. The original formally designed plots are outlined in rocks and blocks.

British-influenced garden landscaping in Jamaica using tropical, mostly nonnative plants traditional to Caribbean landscaping.

public gardens focused on introducing nonnative plants as alternative commercial crops to sugarcane and tobacco so as to expand an island's agricultural, horticultural, and economic diversity. Botanic gardens were also created more recently, such as the Nassau Botanical Gardens developed from a rock quarry.

These imported specimen trees, shrubs, and potential crops formed the backbone of Caribbean public gardens. Such formal-looking gardens were very much part of European cultural history exported into their Caribbean colonies.[33] Depending on the era, the colonial culture, and designer predilections, botanic and estate gardens might be built to be reminiscent of an Italian renaissance garden with its vistas flanked by sharply trimmed hedges and fountains, or a formal, symmetrical jardin á la Françoise with its well-organized flower patches, or an English landscape park's idealizing nature into expansive lawns, lakes, and tree orchards while keeping it all well under control. Not much in these landscaping designs was inherently tropical, acknowledged traditional smallholder agriculture, or incorporated indigenous nature.

As Caribbean islands were exchanged among European nations, garden influences and priorities also changed although gardens tended to be more long-lived than some instances of colonial ownership. For many islands, especially in the eastern Caribbean, eventually England set the pattern where garden history started with royal and noble hunting grounds and estates, evolving into formal gardens, then into wild-design gardens, and finally into publicly accessible gardens. Lancelot Brown created romanticized versions of nature – envisioning scenes had nature got it right in the first place; and William Robertson's wild garden movement evolved into the typical English garden with mixes of pretty and useful plants lining the property.[35, 36] These trends were followed in the estates of the wealthy people who also owned estates in the Caribbean.

Nearby North America also had influence.[37] North American and European garden ideals differed. Europeans for recorded history lived in a managed, cultural landscape, which was reflected in their tidy gardens. North Americans, having conquered their continent's wilderness, attempted to both tame and emulate nature in their gardening. Frederick Law Olmsted and followers created naturalistic-looking parks and gardens, which actually had little relation to the environment that once occupied the space. Nature emulation also became a foundation for the modern American native plant movement. Historically, the Caribbean garden is more European than North American; but recovering Caribbean nature has much in common with contemporary North American native plant and natural gardening movements.

Temperate landscaping in the tropics is maintainable only with great difficulty and against high odds using oft-mowed lawns, seasonally replaced exotic flower beds, regularly trimmed hedges, fertilizers, pesticides, and water-squandering irrigation. Whereas a garden in England may need a bit of trimming once or twice a year and a periodic raking, in the wet tropics plants actually grow and to keep a formal garden in check takes the continuous engagement of teams of cutlasses, rakes, and more recently trimmers and leaf blowers. It is little wonder that such temperate-emulating landscaping over time usually falls into various levels of disrepair.

Over time, Caribbean botanic gardens and some large estates transformed into public parks and, being among the limited public open space, became well appreciated but often heavily trampled features of their communities. In recent decades, some historic public and estate gardens are being reborn as tourist and instructional gardens displaying interpreted plots of heritage plants, utilitarian herbs, backyard vegetables, and native plants. Within a botanic or public garden, the larger the area that can be devoted to native plants and natural garden areas, the better for wildlife. The National Botanic Garden of Cuba devotes 120 hectares to indigenous plants, especially trees, organized around major natural plant communities. The Leon Levy Native Plant Preserve on Eleuthera, The Bahamas, focuses on native plants. Queen Elizabeth II Botanic Park on Grand Cayman emphasizes not only native plants but also endemic animals, including the critically endangered blue rock iguana. Scientific and technical staff of such gardens have the expertise to provide advice to the public and government on conservation programs and locally appropriate uses of native plants in recovering nature. Education programs can provide targeted instruction, outreach, demonstration, and advice advocating for heritage, folkloric, edible, and native plants. Modern botanic gardens can be at the forefront of nature recovery in the Caribbean.[38–43]

The value and use of native plants may be highlighted in nearly any environmental and cultural education program. Beyond botanic gardens, nature centers provide targeted education, outreach, and demonstration on useful and native plants. Schools establish demonstration gardens of edible, butterfly, and native plants tended by students that instill appreciation for gardening, pollinators, and indigenous species. Parks and other areas dedicated to active recreation tend to not be so studiously landscaped as are gardens and nature centers, but these parks can be part of nature recovery if only by replacing nonnative plants with wildlife-friendly native species. There are many opportunities.

North American-influenced, naturalistic garden feature at the Leon Levy Native Plant Preserve in The Bahamas that is meant to look effortlessly natural with native plants but nonetheless is constructed.

3.3 Private gardens and other small areas

Although protected areas and public gardens form the backbone of places for nature recovery, as noted in the Introduction, most Caribbean land is privately owned or privately managed. Official protected areas always will be few because most require large space and all require significant, persistent government and community commitment. Fortunately, state-protected areas and public gardens are but one set of places managed to contribute to a more-natural environment. The rest are smaller gardens, very broadly defined to include private and semi-private home gardens and yards, estates, planned communities, small or large farms, orchards and pastureland, usufructuary property, village commons and squares, pocket parks, urban gardens, commercial sites, or even roadsides. Actually, nature recovery activities tend to be less complicated at such smaller, local scales. Local perspectives can prevail, personal goals count, and they are supported by owners and benefitting users. Recovering nature through creating more natural gardens and other small area plots is a way to overcome the history of colonial-era farming and to ameliorate some effects of subsequent development-focused land use. In the Caribbean, gardening and landscaping more naturally are forms of decolonization.

Designing, establishing, and managing small plots such as gardens in more naturalistic ways have attracted many advocates in the temperate zone.[44–49] These experiences and perspectives are worthwhile for background, keeping in mind that insights being

Garden landscaped with a mix of native plants and wildlife-friendly nonnative plants.

shared are temperate. Recovering nature in Caribbean yards and gardens is not only about breeding-season oak-tree caterpillars for bird food but about sustaining diverse plants that provide a myriad of other invertebrates, fruits, seeds, and flowers accessible to birds and other wildlife throughout the year. And a Caribbean garden needs to supply year-round structure, cover, and security. What is in common between temperate and tropical natural gardening is that both are underpinned by native plants.

Like much of the landscape, Caribbean gardening history is constrained by colonial land-use practices and those that followed. Post plantation-era patterns of small farm development and abandonment tended to leave patches in different stages of succession. For the environment as a whole, such diversity of habitat patches is natural, imitating storm effects. But in more recent history, such patches are seldom allowed to recover to native vegetation. The natural role of patch diversification across the landscape can be taken over by establishing small-scale natural areas, including private natural gardens.

Garden-scale nature recovery relies on perspectives and actions – perhaps changing perspectives and actions – of individuals who own, live, farm, build, or work on a patch of land.[49] For urbanites and suburbanites, appreciating a role for gardens may not come easily as gardens may not obviously seem part of an increasingly urban lifestyle, although they certainly can be, as exemplified by urban garden patches nurtured in many towns. For those engaged in family farming or other community-based agriculture or having farm output expectations, carving out a portion of their allocation for natural rather than agricultural production might be considered personally uneconomical. For homeowners, gardeners, land stewards, and landscapers, reverting some land to more naturalistic plantings may not come easily. The legacy of highly controlled European gardens is hard to shake.[50, 51]

The more native plants in a yard, the better for nature. However, a small-scale landowner or manager no doubt needs to provide for multiple uses of limited space. A private gardener likely may want to have fruit trees; a public garden may want to keep and interpret parts of its historic colonial design and honor its long-standing specimen trees. Gardens and other small plots need not be turned over entirely to recovering nature, a part will do. Adding native plants to existing orchards (bosquet, groves, huerto), hedgerows, or pastures provides diversification for wildlife. However desirable it might be to recover as much of nature as possible, not expecting a garden to be entirely natural is a core concession on the limited land base of the Caribbean. Realistically, few would want their yard composed entirely of scrub or pine forest. And few public gardens would be able to provide multiple functions for the community if comprising untrammeled native forest.

Garden-scale nature recovery has numerous values to the community beyond its borders including providing ecosystem services, adding tree cover, providing habitat for native animals, controlling runoff, enhancing soil fertility, ameliorating heat from artificial structures and roads, reducing noise, creating contemplative and family space, and providing natural products for the family's use. At an island scale, garden-scale recovery is a nature-based solution to protecting water supplies, providing aquifer recharge, fixing carbon, combating climate change, and contributing to sustainable development. Every restored bit of every island's landscape, both public and private, protects the national patrimony of nature. Given that the Caribbean environment has been under human-initiated pressure for as long as humans have occupied the islands, what the individual gardener and landscaper can do with a plot of land is to help relieve this pressure by re-creating natural habitats for native plants and animals. Every little bit helps; as more people add places emulating nature, the island and region become increasingly restored as a whole.

Bahama Palm Shores community intentionally maintaining pinelands and wildlife-friendly elements within a neighborhood development.

3.4 Community regreening

Restorations that are blocks of habitat that function more closely to natural ecosystems provide the most benefits to native plants and animals. However, there is value in simply exchanging nonnative plants for indigenous species. A single specimen in a yard, a mixed-species native hedge, shrubs along the road, and patio and porch containers with animal-friendly plants can positively enhance the environment. Programs and attitudes that advocate and enable such community-wide plantings are often called regreening.

Regreening can also occur on larger scales. This may involve widespread planting of new trees and shrubs, or farmer-managed regreening by regrowing trees and shrubs from existing stumps, roots, and seeds, or community-level tree planting and creating naturalistic thickets. Not all such regreening results in natural ecosystems. Widespread use of invasive, nonnative plants in extensive regreening programs has had much the opposite effect. When regreening goals include creating an agroforestry industry, increasing cropland, or increasing active community land uses, native ecosystems are likely not being recovered. Recovering nature through regreening requires the use of plants indigenous to the area.

Tree-planting programs occur in communities around the world. Community regreening involves planting native trees and shrubs complementary to or as replacements for nonnative ones. One goal is simply to rebuild the tree canopy in a community for all the benefits trees provide. Other goals range from ecological to aesthetic. In addition to trees, native shrubs need to be part of regreening programs, as they provide the core of natural vegetation. Regreening efforts, with local grassroots movements, not only enable restorations but also help change local mindsets, policies, and practices.

3.5 Connectivity

Although protected areas provide a conservation backbone for recovering nature on an island, such bounded areas are necessarily patches within a larger, divergent, culturally determined landscape. So inevitably, parks and protected areas are but fragments of what once was a continuous natural environment. Habitat fragmentation has been widely identified as one of the greatest challenges to recovering nature. The smaller the patch, the fewer species it can contain over a long period. The more isolated the patch, the fewer species will be able to move into it. Moreover, the Caribbean landbase is inherently fragmented among thousands of islands having varying degrees of isolation. Connectivity was identified in Chapter 2 as one of the core attributes of recovering nature.

Ecosystems, although bounded by definition, are not isolated but rather open systems in communication with the environment surrounding them. What happens nearby affects what happens within. Adjacent or nearby incompatible uses can be devastating to a natural area even though protected. Pollution, pesticides, invasive nonnative plants and animals, hydrological diversions, unmanaged fire, poaching, and woodcutting tend

Pearly-eyed thrasher on a garden fence, benefitting from natural gardens' stepping-stone connectivity.

to not respect boundaries. So, protected areas need to be managed to buffer what might be happening nearby.

Within an island, the footprint of restored nature expands with the addition of each small restored place that together form a matrix that overall conserves biodiversity. Connectivity of such natural fragments across an otherwise cultural landscape is required for the maintenance of overall biodiversity, broad-scale ecological and evolutionary processes, and resilience.[52] Some birds can traverse relatively large gaps. Butterflies and other flying animals and those that disperse by wind can also bridge gaps between native habitats. Lizards, snakes, frogs, and turtles can traverse smaller gaps but are more at risk moving through inhospitable habitat. Although seeds dispersed by air can travel long distances, many topical plant seeds do not travel far from the parent unless dispersed by birds or bats. The more connected protected areas are to other natural areas, the greater the biodiversity value of recovered, restored, or regreened sites as a whole. Individual protected areas can be purposely connected by adding corridors of hospitable habitat. Smaller sites, such as gardens, provide stepping stones. Community regreening scatters compatible trees and shrubs across the landscape.

The importance of corridors is well appreciated in the Caribbean.[53–55] Seven large conservation corridors have been identified connecting 47 Key Biodiversity Areas: Massif de la Selle-Jaragua-Bahoruco-Enriquillo Binational Corridor, Haiti and Dominican Republic; Cordillera Central, Dominican Republic; Massif de la Hotte Highlands, Haiti; North Coast Forest-Cockpit Country-Black River-Central Spinal Forest Conservation Corridor, Jamaica; Surrey County Corridor, Jamaica; Iyanola-Castries and Dennery Waterworks Reserve and

Marquis-Mandele Protected Landscape, Saint Lucia; and Saint Vincent Central Mountain Range, Saint Vincent and the Grenadines.[5] These corridors alone contain populations of over 200 globally threatened species, over 100 classified as critically endangered or endangered. An international and inter-island example is the Caribbean Biological Corridor in the Greater Antilles connected by their common marine environments and animal movements, including bird migration. Corridors can connect parks, reserves, Key Biodiversity Areas, Important Bird Areas, Ramsar sites, Western Hemisphere Shorebird Reserve Network (WHSRN) sites, and Alliance for Zero Extinction sites. The Caribbean Biological Corridor incorporates the Key Biodiversity Areas corridors in Haiti and the Dominican Republic. There are certainly opportunities for other corridors, from local to international, throughout the region. Connectivity is a key to restoring Caribbean nature.

References

1. KBA. 2023. "Key Biodiversity Areas." www.keybiodiversityareas.org
2. IUCN. 2016. *A Global Standard for the Identification of Key Biodiversity Areas, Version 1.0, 2016*. Gland: International Union for Conservation of Nature. portals.iucn.org/library/sites/library/files/documents/2016-048.pdf
3. Wege, D. C., and V. Anadón-Irizarry, eds. 2008. *Important Bird Areas in the Caribbean: Key Sites for Conservation*. Cambridge: BirdLife International.
4. Anadón-Irizarry, V., D. C. Wege, A. Upgren, R. Young, B. Boom, Y. M. León, Y. Arias, et al. 2012. "Sites for Priority Biodiversity Conservation in the Caribbean Islands Biodiversity Hotspot." *Journal of Threatened Taxa* 4:2806–44. threatenedtaxa.org/index.php/JoTT/article/view/786/1407
5. CANARI. 2020. "The Caribbean Islands Biodiversity Hotpot." Caribbean Natural Resources Institute. www.cepf.net/sites/default/files/cepf-caribbean-islands-ecosystem-profile-december-2020-english.pdf
6. Wege, D. C., D. Ryan, N. Varty, V. Anadón-Irizarry, and A. Pérez-Leroux. 2010. *Ecosystem Profile: The Caribbean Islands Biodiversity Hotspot*. Washington: Critical Ecosystem Partnership Fund. www.cepf.net/sites/default/files/cepf-caribbean-islands-ecosystem-profile-december-2020-english.pdf
7. Stattersfield, A. J., M. J. Crosby, A. J. Long, and D. C. Wege. 1998. *Endemic Bird Areas of The World: Priorities for Biodiversity Conservation*. Cambridge: BirdLife International.
8. Luther, D., J. Cooper, J. Wong, M. Walker, S. Farinelli, I. Visseren-Hamakers, I. J. Burfield, et al. 2021. "Conservation Actions Benefit the Most Threatened Species: A 13-year Assessment of Alliance for Zero Extinction Species." *Conservation Science and Practice*. conbio.onlinelibrary.wiley.com/doi/full/10.1111/csp2.510
9. AZE. date unknown. "Conserving the World's Most Threatened Species." Alliance for Zero Extinction. zeroextinction.org.
10. UNESCO. date unknown. "Biosphere Reserves in Latin America and the Caribbean." en.unesco.org/biosphere/lac
11. UNESCO. 2023. "World Heritage List." whc.unesco.org/en/list/
12. Ramsar. 2023. "Ramsar Sites Information Service." Convention on Wetlands. rsis.ramsar.org/ris-search/caribbean?pagetab=1
13. Alvarez Malvido, M., C. Lazaro, X. De Lamo, D. Juffe-Bignoli, R. Cao, P. Bueno, C. Sofrony, et al., eds. 2021. "Latin America and the Caribbean Protected Planet Report 2020." RedParques, drive.google.com/file/d/1R4AbdemAH7mUgACuXJ0ovqncuJ8no4AU/vie

14. Protected Planet. 2021. *Protected Planet Report 2020*. Cambridge and Gland: UNEP-WCMC and IUCN. www.protectedplanet.net/en/about
15. World Bank. 2020. "Terrestrial Protected Areas – Latin America & Caribbean." data.worldbank.org/indicator/ER.LND.PTLD.ZS?locations=ZJ
16. Ruiz Plasencia, I. 2016. *Las Áreas Protegidas de Cuba*. Havana: Centro Nacional de Áreas Protegidas.
17. Dudley, N., ed. 2008. *Guidelines for Applying Protected Area Management Categories*. Gland: IUCN. portals.iucn.org/library/sites/library/files/documents/pag-
18. Organisation of Eastern Caribbean States. 2009. "Tropical Forest Ecosystems." In *Biodiversity of the Caribbean*. www.oecs.org/perb_docs/bc_part1a_biobasics.pdf
19. UN Food and Agricultural Organization. 2020. *Global Forest Resources Assessment 2020. Main Report*. Rome: FAO. DOI:10.4060/ca9825en
20. McGinley, K. 2008. "The Caribbean." In F. Cubbage, ed. *Forests and Forestry in the Americas: An Encyclopedia*. Society of American Foresters and International Society of Tropical Foresters. sites.google.com/site/forestryencyclopedia/Home/The%20Caribbean
21. IUCN. 2017. "Green List. Global Standard." iucngreenlist.org/standard/global-standard/
22. Frome, M. 1991. *Regreening the National Parks*. Tucson: University of Arizona Press.
23. Cairns, S., N. Dudley, C. Hall, K. Keeneleyside, and S. Stolton. 2012. *Ecological Restoration for Protected Areas: Principles, Guidelines and Best Practices*. Gland: IUCN. www.iucn.org/resources/publication/ecological-restoration-protected-areas-principles-guidelines-and-best
24. Janishevski, L., G. Santamaria, S. Gidda, H. D. Cooper, and P. H. S. Brancalion. 2015. "Ecosystem Restoration, Protected Areas and Biodiversity Conservation." *Unasylva* 245:19–28.
25. Convention on Biological Diversity. 2023. "Kunming-Montreal Global Biodiversity Framework, 2030. Targets." www.cbd.int/gbf/targets/
26. High Ambition Coalition for Nature and People. 2022. "30x30 High Ambition Coalition Ready to Deliver." www.hacfornatureandpeople.org/home
27. UNEP. 2022. "COP15: Nations Adopt Four Goals, 23 Targets for 2030 in Landmark Biodiversity Agreement." www.cbd.int/article/cop15-cbd-press-release-final-19dec2022
28. United Nations Environment Programme. date unknown. "About the UN Decade." www.decadeonrestoration.org/about-un-decade
29. Gann, G. D., T. McDonald, B. Walder, J. Aronson, C. R. Nelson, J. Jonson, J. G. Hallett, et al. 2019. "International Principles and Standards for the Practice of Ecological Restoration. Second Edition." *Restoration Ecology* S1–S46. DOI:10.1111/rec.13035
30. UNEP. date unknown. "UN Recognizes Initiative to Restore Wealth from the Oceans with Special Award. United Nations Environment Program." www.decadeonrestoration.org/stories/un-recognizes-initiative-restore-wealth-oceans-vanuatu-st-lucia-and-comoros-special-award
31. Coghlan, R. 1975. *Landscape Gardening in the Tropics*. London: Hart-Davis, MacGibbon.
32. Collett, J., and P. Bowe. 1998. *Gardens of the Caribbean*. London: Macmillan Education.
33. Huxley, A. 1998. *An Illustrated History of Gardening*. New York: The Lyons Press.
34. Chambers, N. 2007. *Joseph Banks and the British Museum: The World of Collecting, 1770–1830*. London: Pickering & Chatto.
35. Rutherford, S. 2016. *Capability Brown and His Landscape Gardens*. London: National Trust.
36. Robertson, W., and R. Darke. 2011. *The Wild Garden*. Cambridge: Cambridge University Press.
37. Beveridge, C. E., and P. Rocheleau. 1998. *Frederick Law Olmsted: Designing the American Landscape*. New York: Universe Publishing.

38. Burbridge, B., and P. W. Jackson, eds. 1998. *Conservation Action Plan for Botanic Gardens of the Caribbean Islands*. London: Gardens Conservation International.
39. Maunder, M., A. Leiva, E. Santiago-Valentín, D. W. Stevenson, P. Acevedo-Rodríguez, A. W. Meerow, M. Mejía, et al. 2008. "Plant Conservation in the Caribbean Island Biodiversity Hotspot." *The Botanic Review* 74:197–207. DOI:10.1007/s12229-008-9007-7
40. Francisco-Ortega, J., J. B. Jestrow, and M. P. Griffith. 2019. "Partnerships Between Botanic Gardens and Universities in a Changing Caribbean World." *Hemisphere* 28(1):Article 1. digitalcommons.fiu.edu/lacc_hemisphere/vol28/iss1/1
41. Acosta Cruz, M., L. Mugica Valdés, and K. Aguilar Mugica. 2022. *Aves del Jardin Botanico Nacional de Cuba*. Havana: Sello Editorial AMA.
42. Monter Espin, R., L. Mugica Valdés, M. Acosta Cruz, K. Aguilar Mugica, R. Simon Vallejo, and J. E. Martínez Gonzalez. In press. "Botanic Gardens and their Role in the Study and Conservation of Cuban Birds." *Revista del Jardín Botánico Nacional*.
43. Kushlan, J. A., and K. Hines. 2014. *Birds of Fairchild*. Morley: Acclaim Press.
44. Druse, K. 1988. *The Natural Garden*. New York: Clarkson N. Potter, Inc.
45. Tallamy, D. W. 2009. *Bringing Nature Home: How You Can Sustain Wildlife With Native Plants*. Portland: Timber Press.
46. Matloff, G., C. Bangs, and L. Johnson. 2010. *Paradise Regained: The Regreening of Earth*. New York: Copernicus Books.
47. Weaner, L., and T. Christopher. 2016. *Garden Revolution: How Our Landscapes Can Be a Source of Environmental Change*. Portland: Timber Press.
48. Richards, M. A. 2018. *Regreening the Built Environment: Nature, Green Space, and Sustainability*. London: Routledge.
49. Muller, A., and M. Bourne. 2021. *Regreening the Globe Depends on Grassroots Movements*. Cologny: World Economic Forum. www.weforum.org/agenda/2021/01/regreening-the-globe-depends-on-grassroots-movements-how-best-to-help-them-grow
50. Webster, A. 1986. *Caribbean Gardening,* 3rd edition. Kingston: Novelty Trading Company.
51. Bannochie, I., and M. Light. 1993. *Gardening in the Caribbean*. Oxford: Macmillan Education.
52. Soulé, M., and J. Terborgh. 1999. *Continental Conservation: Scientific Foundations of Regional Reserve Networks*. Washington: Island Press.
53. Conservation Corridor. 2023. "Connecting Science to Conservation." conservationcorridor.org
54. CBC. 2022. "Corredor Biologico en el Caribe." cbcbio.org
55. UNEP. 2022. Gerhartz-Muro, J. L., and N. Vina-Dávila, compilers. "Project to Strengthen the Biological Corridor in the Caribbean: Strategic Plan for the Biological Corridor in the Caribbean 2022–2030." cbcbio.org/wp-content/uploads/2023/04/CBC-Strategic-Plan-EN.pdf

4 Practical matters

The practicalities of recovering Caribbean nature differ from typical land management and many usual gardening and landscaping practices, especially temperate-derived practices. Ecosystem restoration in large protected areas may differ in magnitude and complexity from relandscaping a public garden, regreening a community, or planting native trees around a commercial building or at a resort; but, they have far more in common than not. The same considerations need to be evaluated such as plant choice, design and management, the impacts of rain and storms, fertilizing and pests, and ongoing care.

4.1 Plant choice

Recovering nature's fundamental methodology, at whatever scale, is the use of indigenous plants. Large-scale ecosystem restorations aim to achieve as complete a complement as possible of site-appropriate native plants. Smaller-scale, space-limited restorations must use fewer species so plant choice needs to be thoughtful but need not be overwhelming as a few native plants can attract animals that may then introduce other appropriate native plants.

In some cases, identifying a plant is easy – everybody knows it, at least by its local name. Other times, it is hard as many native tropical plants look alike, especially their leaves; and there is a scarcity of well-illustrated, complete guides for the region. Sometimes even experienced people are unsure. Plant species are still being discovered, including unraveling the relationships among similar species from different islands. Identification usually involves a combination of leaves, trunks, flowers, fruits, structure, and location. As scientific species descriptions use a specialized and rather intimidating vocabulary, books and websites with photos and simpler descriptions tend to be useful. Chapter 8 provides more information on such resources. Increasingly, websites and applications can assist, although plant identification apps may take a while to be entirely reliable in a diverse, endemic-rich, and continuously restudied tropical flora such as the Caribbean. Local plant experts, government agencies, conservation organizations, hobbyists and garden clubs, and staff of botanical gardens, plant nurseries, and nature centers often are able to share their knowledge.

In selecting plants, it is useful to consider the flowering and fruiting seasons of each so as to be sure to have flowers, insects, fruits, and seeds available year-round. Hummingbirds, bananaquits, and pollinating insects need nectar-bearing flowers year-round. Plants that attract or host insects and other invertebrates are particularly valuable during breeding seasons of birds, lizards, and other insect eaters. Year-round, Caribbean birds

DOI: 10.1201/9781003396321-5

Selection of native plants used to landscape a restaurant terrace.

need fruits and seeds, their most-used food overall. Even winter resident and migrating birds that are primarily insectivorous in summer in North America eat fruits during their time in the Caribbean, often spent in gardens and scrub-type habitats. Excellent wildlife plant species may provide all of the above, but a collection of species provides a diversity of food.

Sources for plants include plant nurseries, agricultural supply stores, garden shops, random stores carrying plants, private gardeners, garden clubs, naturalist and conservation societies, botanical gardens, environmental centers, forestry and agriculture departments, and neighbors. Unfortunately, so far, demand for native wildlife-supportive plants has been insufficient to inspire most plant nurseries to carry much in the way of native plants. The larger and more prosperous the island and the more into gardening and conservation, the more likely there will be suitable native plants for sale somewhere.

Many native plant species, including those recommended in this book, are widespread in the Caribbean and can be imported from other islands. Caribbean plants are also grown in Florida, which shares much the same native flora as nearby limestone islands, and are available elsewhere in the hemispheric horticultural trade. But, using imported plants is not to be done lightly as too little is known about genetic distinctiveness of plants from around the Caribbean; using stock foreign to an island might affect the genetics of otherwise isolated indigenous plant populations. Furthermore, island biosecurity is an issue as introduced plants can bring pests, disease, and nonnative animals. They can also affect indigenous populations and other species. It is best, to the extent possible, to find plant stock from the island where it will be used. Demand may create a market to encourage nurseries, organizations, and individuals to propagate and market environment-friendly, native, local plants, to encourage government agencies to make environment-beneficial

Native plants, including cocoplum, buttonwood, and firebush, for sale at a plant nursery in Nassau, The Bahamas.

plants available, and to support charities that advocate for the environment in sponsoring native plant distributions. Supporting native plant societies, environmental centers, and botanic gardens enhances communities and networks of mutually supportive native plant practitioners.

Rather than being purchased, plants can be propagated from seeds, cuttings, division, transplanting seedlings, a bit more complicatedly by air layering, or by more than one of these techniques. Chapter 5 offers guidance for how recommended plants might be propagated. A great advantage of propagation is that the plants are local. In general, plants, especially trees, grown from seeds develop more extensive and deeper root systems than those propagated in other ways. Nonetheless, other methods are also effective and usually quicker. Although taking time and effort to get started, once established, Caribbean shrubs and trees native to a location will generally grow quickly and are well adapted to local soils and local wet-season–dry-season cycles.

Many tropical seeds, unlike temperate plants, do not last long after maturity and need to be processed quickly, although there are exceptions such as some palms. In getting some seeds to germinate, imitating nature is best. Many of the plants recommended in this book have fruit that birds eat. Passing through a bird's acidic gut, the seed coat is prepared for germination, a process approximated by soaking seeds for a short time in water or weak acid or rubbing with sandpaper, allowing seeds to hydrate. Generally, it is best to plant seeds in flats so that they can be monitored, protected, and managed until transplanted. It also is best to start plants in soil like that of their natural habitat, as a test of suitability. Some seed stock may be available in stores, online, or with necessary permissions collected from the bush, neighbors' yards, or public plantings. They also can be preprocessed by birds, secured by gathering sprouts under roost and resting sites; resting sites such as rails and rope lines can be provided for the purpose.

With permission and, where legal, plants can be transplanted from the wild. There may be regulations overall or in specific locations or it may require permits. Seed collection may be permitted even when adult plants are protected. Many of the native plants this book recommends are naturally weedy and can be obtained from disturbed sites. With permission, native plants can be rescued from lots pending development. Transplanting small seedlings, especially species with tap roots, tends to be more successful than digging up larger specimens. Palms and xeric species are exceptions in that they transplant well. Scrub, forests, shore, pinelands, wetlands, overgrown lots, and the roadside can be sources for plant stock. In the temperate zone, transplanting is best done when plants are dormant in winter. In the Caribbean, what dormancy occurs is during the dry-season period of water stress; transplanting in the wet season lets rainfall take care of irrigation during establishment.

Many tropical plants can be propagated by cuttings of stems, roots, or leaf buds. These are often the fastest and most successful ways of propagation, and local genetics is more assured. There are many sources for detailed instructions on propagation techniques. It is not as daunting as it might seem. Some plant stems have dormant buds that root on their own; others need to have stems nicked a bit to inspire root growth. Some species benefit from rooting hormone application. It is best to stick cuttings first in a pot rather than in the ground; after roots and leaves develop, they can be transplanted. Trial and error are inevitable. These techniques create clones and so do not contribute to maintaining the genetic diversity of the population as seed propagation might.

Native plants being propagated from cuttings.

4.2 Design and management

Designing, planning, implementing, and managing nature recovery at various scales can be as complicated as restoring a large forest ecosystem or as simple as planting a few native trees. At the large-area end of the spectrum are programs of ecosystem restoration and reforestation, backed by an extensive body of research and experience.[1–8] Tropical forestry practice, forest restoration, and tree-planting programs are disciplines and practices of long-standing with extensive professional engagement, support of international agreements, impetus from climate change concerns, and national and multinational programs that are planting millions of trees. The most extensive example of a multinational program may be the Great Green Wall initiative in Africa and the Nobel Prize-recognized Green Belt Movement of Wangari Maathai. At the smaller end of the spectrum are design decisions about establishing areas of natural plantings in public and private gardens and commercial settings. At this scale too, there is a substantial and useful literature on landscaping and natural gardening practices, although primarily from a temperate perspective and, for the Caribbean, mostly focusing on herb, edible, and medicinal gardens.[9–24]

Given the extensive background available on forest practices, ecosystem restoration, landscaping, and gardening, there is no need to review the fields here. It is, however, worthwhile to consider design and management specifically in the context of recovering Caribbean nature. From parks to gardens, decisions need to be made about what to do, when to do it, and how to manage in the long term. Given all the high-level, global organizing of protected area conservation, it is sometimes easy to overlook the importance of local restoration teams. Much work on Caribbean protected areas is funded by international agencies that do not provide perpetual support. Local conservation organizations can be effective long-term.

A natural area, big or small, needs to have a boundary that everyone recognizes, unambiguously defined, platted if possible, marked, and enforced. Although rather obvious, this actually is fundamental; without a boundary, conflicts with other interests are inevitable. Incompatible activity can devastate a nature recovery project. For small areas, signs, fences, and hedging create recognizable boundaries. For large areas, boundary defense may require not only boundary marking but also reliable funding for guards or rangers; and funding is seldom easy.

Reforestation and afforestation require special consideration as they involve commercialization. Cartesian-spaced, monoculture tree plantations provide limited diversity and wildlife value yet are tallied as forest coverage. Reforestation tends to mean planting commercially valuable, usually nonnative species in ways that facilitate harvest rather than recovering nature. Interplanting native trees and shrubs or harvestable shade plants, such as coffee or cacao, within agroforest increase structural diversity and food sources. Interplanting a diversity of noncommercial, wildlife-friendly shrubs would do more. Retaining or recovering a full measure of an area's nature requires native plant diversity.

When a forest, or other protected area, lacks the full complement of plants native to the site, introducing missing or poorly represented indigenous species may be part of a restoration plan. Few restorations return all species; planting native plants as available is a practical approach. Visitor and headquarters areas can be planted with indigenous plants,

An example of a large reserve, the Tobago Cays Marine Park in the Grenadines protects habitats such as this coastal scrub.

which serve educational purposes and also reduce maintenance. Demonstration plots can highlight native wildlife-friendly and indigenous medicinal plants. Butterfly and hummingbird gardens take little space but have high impact for animals and for education. Citizen projects that actively remove nonnative plants not only benefit native plants and animals but also send a message as to their value.

Nature recovery needs to be planned. For large areas, authorities undertake significant planning and assessments documenting the state of the resource, intended sustainable uses, desired end-state and trajectory measured against benchmarks, and mechanisms for adaptive changes in management based on experience and shifting conditions. Considerations might include the continuation of ethnobotanical or other cultural traditions, environmental justice concerns, visitor use, or sustainable use of resources. Other considerations might be economic, cultural, philosophical, educational, and sociological. The involvement of local residents, users, specific support groups, not-for-profit organizations, scientists, naturalists, and other stakeholders provides knowledge, insight, and support for planning decisions. The end product for large-scale restorations is a definitive plan to achieve stated goals through an ongoing management process, preferably that over time will result in the restoration becoming more self-supporting.

Planning for small areas should be less complicated but still benefits from design. Formal principles of landscape design can be found in many books; traditional elements include the concepts of unity, scale, balance, simplicity, variety, emphasis, and sequence.[25,26] These concepts can suggest how a garden or yard or larger landscape can be designed for aesthetics as well as for the various uses it is to have. However, for small personal plots, it need not be that complicated. Sometimes simpler intuition, practicality, and personal taste go far. Beyond choosing native plants, the natural area or garden should be designed to be as naturalistic, ecological, diverse, and as connected as possible and take into consideration local land form, soils, runoff, multiple uses, costs, and maintenance. The overall landscape design can be messy or tidy, crowded or delineated, functional or aesthetic, maintenance-heavy or maintenance-light. However, more widely acceptable natural landscaping is neat, yard-sized, with nectar-bearing flowers and small fruit.

Accommodating to the size, shape, and conditions of a site is a primary concern in landscaping with nature. Reserved areas, public parks, and estates may have many hectares or even square kilometers to work with. Subdivision lots also may have considerable space; but even an average home lot of a few hundred square meters has space. A 20 × 35 m^2 lot covers 700 square meters. One can scale down further all the way to a container with plants used by birds and butterflies. There are constraints, of course. Public parks, family land, community gardens, commercial lots, and government allotments have competing requirements for their use. Zoning, covenants, and neighborhood opinion may govern private land.

The usual landscaping advice is to first concretize a conceptual design by conducting an inventory of a property marking off ideas in real space and then create a plan on paper for what the plot is intended to become. An advantage of writing it down is that it gets everyone on the same page, literally. A landscape architect's plan would be quite complete. But, for modest-sized private natural landscaping, the base map need not be so definitive. A simple conceptual diagram having more or less the right dimensions and noting the landscape plans to be implemented might do. And sometimes the design is just mental. However, deciding aspects of a landscaping design in advance is a lot easier than re-digging.

Another early design step is sorting out how people will use the space. Throughout the tropics, and the Caribbean is no different, the yard is an extension of the house, a place for eating, having an edible or herb garden, drinking, contemplating, sitting, playing,

Well-designed gardens incorporate people-space within planted areas.

Native trees clustered together provide the most environmental benefits including natural habitat structure and storm resistance, as well as shade for pedestrians on this sidewalk.

fixing things, liming, and more. Such activities can take place on the dirt under a wide-spreading, sticky, nonnative banyan tree but are much enhanced within a more-natural garden setting. People-space need not consume the yard; various activities can use the same space. Planning people-space early is crucial because the garden will be trampled to achieve it anyway. Defining people-space becomes even more crucial in public parks and even in large reserves, as these are inherently multiple-use areas.

Where do the plants go? Ideally, native and useful plants occupy any space not taken up by buildings and essential people activities. If starting with a blank plot, options are broad. It is useful to pretend one has an empty lot so as to imagine its possibilities. But most projects contend with existing buildings and other infrastructure. Public parks are constrained by needing to consider multiple constituencies. Any project is constrained by time and money. Nonetheless, much is changeable. Is the big, old, nonnative, densely shading banyan tree really what is wanted and where? People want shade, but butterfly plants need sun. Is there a compromise? Are fences and hedgerows where they perform best? Where can plants take the place of pebbles and dust? Can containers and raised beds expand planting opportunities? Are porches or second-floor balconies available? Where can vines be fit in? Do cars really need to park all over the front yard? There likely is a lot more plantable space than apparent at first look.

Trees also are among early considerations, given how they dominate and determine the structure of a landscape, forest, or natural garden. Ecosystem restorations choose trees to re-create the native forested environment; but as noted previously in smaller patches, there can be but few trees so selection is critical. Trees may already exist to constrain landscape design. Native trees are best situated within patches of diverse forest rather than as individual specimens scattered about the yard. It is in this configuration that trees serve their most complete ecological functions. Single-specimen trees are more like garden statues, to be admired but not used. A natural garden can be overplanted with native trees initially as they will sort themselves out as they compete for space and sun.

Trees provide shade, which can be helpful or not. Awnings and canopies also provide shade; but trees can do it more naturally, more cheaply, more pleasantly, and support animals. They ameliorate heat, sun, and glare and fend off rain showers. A tree with small leaves, wide branches, and perhaps partially winter-season deciduousness provides mottled light that allows shade-tolerant plants beneath. When situated to windward, especially near the coast, trees provide a shield from onshore winds. Situated on the west of the property, they deflect the heat of the afternoon sun. Given the threat of storms, large tree branches should not hover over the roof, and tall trees need to be far enough from structures not to topple on them. Small trees and shrubs may be more desirable to plant near buildings.

Although rainforest species in the proper setting can be gigantic, stretching to 30 m or more, many Caribbean tree species end up much shorter. Size is relative – in the Caribbean, scotch bonnets are sometimes called pepper trees. So, trees in natural areas need not be tall. Many native tree species, especially those found in second growth or on the coast, take to being pruned naturally by wind and salt, and so they are successfully pruned in a garden setting. Except in arid and cool mountainous areas, in the Caribbean, trees can grow fast. It is said in the temperate zone that one plants trees for the next generation. In the Caribbean, the plantsman can enjoy the result. From a small transplant or cutting, some species, gumbo limbo for instance, can be a significant tree in just a few years. Even some tropical species identified in the literature as slow-growers in nature actually grow more quickly under gardening conditions, lignum vitae and hog cabbage palms for example.

A natural area in public parks and private gardens can be part of a community effort to increase and diversify tree cover. Overall, a societal goal for a village, city, or island – with consideration for hurricanes – might be to increase the overall tree canopy using native species. This not only benefits native plant and animal conservation but also moderates high temperatures, reduces runoff, increases water quality, and improves air quality. Research in North America has suggested that tree canopies in urban areas reduced stress, increased public safety, and increased property values and marketability.

Shrubs also need early consideration in planning natural landscaping. Shrubs rather than trees create the bulk of the natural garden, the tier that wildlife and people most interact with. Shrubs provide the structure, color, texture, and much of the wildlife benefit of a natural planting through their shelter, as well as their fruits, flowers, and insects. Mixed-species shrubs form the edge of natural forests and, similarly, edges of natural garden patches. In mixed-species hedgerows, shrubs create wildlife habitat. Hedges are an expected part of a formal European garden, highly manicured, sharply boxed, single species, perhaps crafted into topiary creations, all of which reduce any ecological value. But in the Caribbean, mixed native species hedges or hedgerows are more diverse in all wildlife-enhancing properties and need far less manicuring. Living fences are practical hedges made of closely planted trees that grow from limb cuttings, such as gumbo limbo, or in arid areas, from cactus. Diverse hedges can serve as visual barriers, edges, frames, and as actual barriers, if thick and/or thorny, especially if also entangled with thorny vines. Vines are often overlooked as components of a natural garden. Climbers spread within

A diverse hedgerow of native plants including palms, shrubs, and trees.

Living fence of cactus on Bonaire at Washington Slagbaai National Park.

A diverse lawn in Abaco, The Bahamas, being used by a mockingbird.

the canopy of a natural area filling voids, adding structure, and providing cover for birds and other wildlife. Many have bird-friendly flowers and fruits. They contribute to structure and complexity.

Plants that are relatively short, perhaps under 1 m tall, also provide structure and cover for animals in various ways. These short-statured plants may be herbaceous or woody, short-lived or perennial. Ground covers and beds of short-statured plants are quite traditional in temperate gardens where store-bought bedding plants are changed out seasonally. Native short-statured plants naturally grow along the edges of forest and scrub or along roads and weedy places; they create borders to natural gardens and need not be changed out seasonally. These otherwise weedy, short-statured, native plants are perfect for natural gardens and other plantings.

Natural areas and gardens have little need for typical lawns. The detriments of sod and benefits of a diverse lawn have become well understood in temperate areas; the negatives are even greater in the tropics. A vast expanse of a single-species grass lawn is not a tropical thing; it is a temperate thing. Sod grasses love England and New England, not so much summer heat and humidity, insects and fungus, and a long dry season. Because it is so unnatural and demanding, a monoculture lawn takes a huge amount of maintenance including water, fertilizer, mowing, tilling, aerating, overseeding, sod replacement, soil amendments, insecticides, fungicides, and herbicides – over and over again. And it needs sun, not the shade created by natural habitat. Such monoculture lawns, especially when maintained by bird-food-killing pesticides, are of limited value to wildlife.

Gardens need not be without lawn space. Where needed in a natural garden, a diverse lawn can be created, essentially a short, natural meadow, a patch of many species of low-growing grasses, sedges, herbs, vines, and assorted seedlings. A diverse lawn will create itself from volunteers but can be enhanced with light dustings of mixed-species seed, sod plugs, seeds from wild flowers, and transplants from roadsides and fields. A diverse lawn supports birds, butterflies, and other animals year-round with minimal care.

Epiphytes benefit a natural garden. The Caribbean has a wealth of indigenous epiphytic bromeliads, orchids, and ferns. Epiphytes increase manyfold the habitat space available for communities of insects and other small invertebrates. Installing native epiphytic species in a garden, so long as obtaining them is legal, helps enhance their conservation. Vase-shaped epiphytic bromeliads hold water in which a community of invertebrates and vertebrates such as frogs can live and can provide a water source for animals in the dry season. Both orchids and bromeliads can have impressive floral displays and so bring color and texture to the garden.

4.3 Soil

Soil, described in Chapter 1, is a key to recovering natural areas; but, the pervasive and lingering impact of past agricultural practices in the Caribbean makes understanding contemporary soils and what plantings they can support complicated. How to deal with this region-wide complexity? Fortunately, one has to deal only with a single plot of ground. It is sometimes worthwhile and often free to have soil tested to determine exactly its chemical nature. It also is relatively easy to find out in general about local soil as it is the same as the neighbor's and understood by local plant nurseries, garden shops, and agricultural extension services. Details of previous land use may prove useful. But whatever soil is there, is there. In tropical gardening, it is best to use what is there rather than trying to amend it into something else. Native plants found in the area tolerate current soils, whatever they might be.

Soil provides a large, although mostly unidentified, portion of the Caribbean's biodiversity. Communities of ants, bacteria, beetles, centipedes, isopods, fungi, millipedes, mites, protozoa, slugs, snails, spiders, termites, and worms of various sorts provide ecological linkages, food webs, and pathways of energy and nutrient flow within and beyond the soil ecosystem. These communities of detritivores and saprophytes decompose organic matter. They include fungi and bacteria that have symbiotic relationships with plant roots. Caring for whatever soil there is, and the organisms contained within, is essential.

What is to be done with Caribbean soil in efforts to recover nature? The less the better; and this is a fundamental difference between agriculture and recovering natural areas. Agricultural lands likely need continuing soil management to restore and maintain the fertility that crops extract. Recovering nature is another matter. For large-scale restoration, attempting to reform soil clearly is impractical. But, the same holds for smaller plots too. This might mean accepting that the existing soil will produce an ecosystem different from what historically existed on native soils and choosing native plant species that work in existing altered soils.

Soil organic matter is certainly a big deal in the Caribbean as most soils either have little naturally or what they once had long ago has been lost. One approach widely used to increase organic content of soil is to cover the ground with thick, mostly store-bought, mulch. Mulch also is meant to control erosion, hold water, kill emerging weeds, fend off string trimmers, and look tidy, all of which it does. However, unlike in a temperate deciduous forest, in much of the tropics, such thick mulch seldom accumulates naturally. Instead, here trees have shallow roots and symbiotic relationships with soil mycorrhizal fungi so as to rapidly process the limited nutrients remaining in natural leaf litter as it quickly disappears. That symbiotic fungi are adapted to natural litter and humus, not necessarily to foreign mulch, is not often appreciated. Thick mulch in the tropics often holds too much dampness leading to nonsymbiotic fungi taking over. Some store-bought mulch, such as eucalyptus, inherently repels the insects that otherwise would provide for birds and other animals. Heavily applied store-bought mulch is not very natural in the Caribbean.

The natural sources of organic matter are fallen leaves, flowers, fruits, twigs, and branches of the trees and shrubs themselves. These fall within the canopy area and root zone of the plant that produced them. To be recycled, they are allowed to remain where they fall. Only if adversely intruding on another garden area or pathway might they need to be moved to another area or composted. They definitely should not be sent away or burnt. This sparce, naturally derived litter supports the soil ecosystem and allows ground-feeding birds and other animals access to the invertebrates and seeds found there. Litter coverage should be incomplete with intermittent patches of ground that open the soil to foraging animals. Mulch does look good to those aspiring to a tidy aesthetic. If foreign mulch is considered an aesthetic necessity, it can be confined to decorative roles such as narrow bands along walks and other formal edges. Mulch also can be quite useful in highly managed settings such as edible and herb gardens and to retain moisture to help

Natural mulch in a naturalistic public park composed of leaves and sticks.

newly installed plants get established. In the Caribbean, mulch is to be used sparingly, purposefully, ecologically, and not inflicted on native trees and shrubs.

Unplanted areas are rather common in Caribbean yards, given the difficulty of growing lawns. This can be unattractive and is certainly unproductive. Front yards may be broomed daily leaving behind sterile pavement-like soil. Valuable organic matter of leaves, twigs, dead flowers, and dropped fruits is whisked away into a garbage bin rather than being allowed to remain to form litter. Leaf blowers assure no unauthorized leaves lay about. Swept-clean ground is dusty when dry and muddy when wet. It compacts and affects drainage, and plants cannot grow there. It is so much better functionally and ecologically to plant the yard and leave the litter under the shrubs.

Soils, mulch, and litter lead to consideration of fertilizer, meant to supplement the nutrients available in existing soil and its litter. Although fertilizing, either chemical or organic, is usually needed for most commercial agriculture in the Caribbean, native plants do not need artificial fertilizer and may do poorly when fertilized, altering growth patterns or blooming cycles. Artificial fertilizing runs the risk of changing soil chemistry to the disadvantage of plants, soil fauna, and symbiotic bacteria and fungi. Artificially fertilizing native plants also is a waste of money. Native plants get their nutrients from litter, rain, and the air.

There is another form of fertilizer, much praised in organic gardening, compost. Compost adds organic matter and releases nutrients more slowly than does chemical fertilizer. It takes a lot of biodegradable waste to make a little compost. To some extent, composting may be as much an environmental statement as a gardening tool, providing a way to recycle otherwise discardable organic matter such as kitchen waste, lawn cuttings, sticks, and paper products. Composting takes 3 stages: a short medium-temperature phase, a high-temperature phase, and a maturation phase. There is much to read about composting, mostly from a temperate perspective. In the tropics, composting goes fast and is easier than in cooler climates as everything more or less takes care of itself, quickly. While native plants do not need it, compost is useful in specialty gardens.

A more intense form of natural fertilization available throughout the region is manure, which has been crucial in agriculture for millennia and continues to be a valued local source of plant nutrients and organic matter. In the volume that might be available from horse stables, goat and sheep pens, or chicken runs, fresh manure is easy to overdo. It first needs to mature, decomposing in much the same steps as compost. The high-temperature phase is especially important to kill potentially pathogenic microorganisms. During this phase, nitrogen levels remain too high for plants. By end of the third stage, nitrogen levels have relatively decreased and phosphorus and potassium levels have relatively increased, allowing it to be used as a supplement mixed with compost or natural mulch. Like other fertilizing and organic matter additions, it is not needed by native plants.

4.4 Temperature and rain

The Caribbean climate – temperature and rainfall – has been discussed previously. As noted, even in the tropics, temperature differs from place to place and perhaps season

A summer rainstorm in The Bahamas.

to season. Accommodating to the local temperature regime at macro- and micro-scales needs to be taken into consideration in planning a natural garden. It should influence both plant choice and garden design. Temperature regimes differ among islands, among altitudes, and within islands among sites. Towns and cities tend to be hotter than the countryside. The coast is more moderate than the interior. Even within a single area there are microclimates. Walls, buildings, stone, and asphalt produce warmer spots. Plant shade produces cooler sites. As a generality, an important consequence of the generally warm Caribbean climate is that there is limited seasonal thermal check on the growth of plants, survival of insects, and spread of fungus, bacteria, and viruses. So, plants need to deal with continuous natural herbivory and disease challenges. Native plants can cope with their indigenous herbivores and diseases or they would not have survived them.

Rain similarly differs from place to place and season to season. In the dry season, water becomes precious, likely too precious to be shared with most plants. This is especially the case on small or arid islands; but most islands face a water deficit towards the middle to end of the dry season with wetlands and shallow ponds drying, rivers slowing, and aquifers falling. And in the dry season, residual surface or groundwater that might be available is prioritized for human and agricultural use. Using plants indigenous to a locale makes things easier as these plants will have adaptations to manage through their normal dry season. They may become semi-dormant and deciduous, store water in the trunk, have deeply penetrating roots, or use water-saving photosynthetic pathways. Most islands have native xeric-adapted plants that can be used in restorations. Native terrestrial species do poorly if given too much water in the dry season as flowering cycles may depend on a previous dry season. Nonnative plants, however, lacking dry-season adaptations, likely need supplemental water. Specialty gardens, such as herb and edible gardens, also need supplemental dry-season water.

How to get water in the dry season? One way is to save it from the wet season. Rainwater is free, starts off relatively pure, and can be stored. There is an extensive literature on how to harvest water and use it with ecological sensitivity.[27–31] But, the most efficient approach remains rooftop catchment and cistern storage. There is no more Caribbean approach to water management; over a half million people in the Caribbean use this method for drinking water. The Bahamas, Turks and Caicos, and Saint Lucia are examples of islands employing significant rainwater harvesting. Bermuda, although outside the geographic scope of this book, has traditional architecture that enhances rooftop water capture. Cistern and other captured water can be shared with plants when they need it. Where no longer in domestic use, old cisterns can be repurposed for irrigation.

Typically, rain is collected on a roof or similar chemically inert elevated surface and transmitted by gutters, downspouts, and pipes to cisterns, above-ground tanks, and rain barrels. Personal wells and public water supplies can also be used to fill cisterns and other private reservoirs. Overflow can be directed to a soakaway then to a damp garden with semi-aquatic and aquatic plants, often called a rain garden. Dasheen (taro) does well in rain gardens. Gray water from sinks, showers, and such, depending on local rules, might also be sent to the rain garden. After the rain garden treats it, the water can be channeled to specialty gardens. Even a kitchen sink can provide dry-season water for a vegetable or herb garden.

Roof designed for rainwater capture, Bermuda.

Plants can also be watered from the cistern directly. Below-ground cisterns use a small pump, which can be solar, to distribute the water. Water delivery from above-ground storage can use simple taps and gravity flow. Ideally, a battery-powered timer turns the delivery on and off at the appropriate irrigation time, which in the tropics is in the early morning. Care needs to be taken in using municipal water, as it is, or should be, chlorinated, which can adversely affect plants. An advantage of cisterns and other private reservoirs is that any chlorine has time to dissipate during storage. Water use for small specialty gardens can be modest; and, as noted above, native plantings should not be watered after establishment.

The most useful component of a water distribution system, where available, is drip irrigation. Of course, plants can be watered by hand or hose, but drip irrigation manages timing and amount precisely and automatically. Water passes from a reservoir or other source through small tubes to emitters placed where they are needed. Flow is controlled by the emitters; schedule and duration are controlled by a battery-powered timer at the faucet. Water loss is minimized compared to sprinklers or other aerial applications.

The rainy season may be more complicated than the dry. Rain is variable, sometimes too much, sometimes too little, not often just right. Even normal wet seasons may produce rain daily or over several days causing runoff, puddling, and erosion, especially in culturally determined landscapes lacking natural impediments to flow such as native plant communities. Wet seasons not only support plant growth but also bring high humidity, waterlogged soil, erosion, fungi, and insects. The amounts of rain seasonally falling on windward coasts and montane rainforests can be prodigious. On some islands, impressive rivers carry water downstream into lakes or marshes and to the sea. Other rivers arise in

Drip irrigation provides water in regulated small doses delivered through narrow tubes, shown here.

the lowlands and flow more gently to the sea, perhaps only seasonally or in storms. Water also percolates into the ground to be captured in aquifers, which on some islands float above salty underground water.

In storms, there can be too much water, too fast, which needs to be managed to the extent practicable. The greater the slope, the more extensive and structured are the water management systems that might be needed, typically requiring professional guidance. Lesser sloping lands such as down-slope parks and gardens are affected too as they are the recipients of upgradient water. All this water needs to go somewhere. Down-slope conveyance and retention may be desirable even for more modest rains. Water can be directed by ditches, swales, berms, channels, culverts, and pipes. On steeper slopes, cross-slope swales intercept sheet flow and prevent slippage; pipes and culverts and channels can move water in ways that reduce damage. On shallower slopes, the water can be encouraged into depression and ponds where it can accumulate, be used, or absorbed to replenish the aquifer.

Hurricanes pose even greater challenges. A hurricane, particularly a slow-moving one, can produce hundreds of millimeters of rain in a few days causing flooding, sometimes catastrophic, and landslides. Even less-fierce tropical storms, squalls, and continental frontal systems can bring high wind and rain. Winds strip leaves, snap and twist branches and trunks, send tree limb projectiles flying, uproot trees crushing neighboring plants, create large windthrows, and spread salt spray. Even a mild storm can quickly turn a garden into a mushy trash pile. Hurricanes happen; they always have. From one generation to the next, humans try to forget they are coming again.

Many Caribbean native plants have specific adaptations to survive hurricanes, such as readily losing leaves or branches, regrowth from severed branches and downed trunks,

A tributary to the Buff Bay River, Jamaica, where rocks are being used to direct water flow.

brief salt-tolerance, rapid post-storm leafing, flowering, fruiting, and post-storm sprouting of seeds buried in the litter or newly arrived to canopy gaps. Wild coffees recover and fruit quickly after a hurricane. Bayberry holds onto its fruit. The short stature of many scrub trees make them less affected by storms than tall forest tress. Hurricanes can cause widespread devastation of montane rainforests, taking years to recover. However, after such devastation by hurricane Maria in Dominica, although over three-quarters of the rain forest trees were damaged, tree mortality turned out to be only 10%.[32] Full recovery still will take decades.

There are well-known outdoor hurricane-preparation tasks – best of course taken care of before any storm is approaching – including pruning heavy tree branches over the house, making sure surface-water management systems are functioning, securing outdoor equipment, harvesting coconuts, setting aside water, and maybe getting a generator. Defying convention, trees and palms away from structures need not be overly trimmed, as they can do valuable service as wind breaks. They can be trimmed after the storm. Shrub hedgerows too are windbreaks. Seagrape as part of a hedgerow, for example, is particularly good as a windbreak as it is adapted to windy shores. Native trees and shrubs gathered in a cluster with intertwining roots and branches provide hurricane-resistance. A tree out on its own is likely a downed tree, especially shallow-rooted, nonnative plants such as Australian pines.

The experience post-hurricane is always heartbreaking. The post-hurricane scene is also a restoration program's or natural gardener's heartbreak. Other than hacking access to houses and removing trees from electric lines, vegetation is not usually a top post-storm priority. Pressing necessities of electricity, water, health concerns, transportation, and structural and equipment damage need to take precedence. Nonetheless, plants benefit post-storm from rinsing salt from leaves and soil, trimming broken stems and branches, quickly re-standing fallen trees, draining water to inhibit the post-storm mosquito plague, repairing flood effects, and adding to the accumulating mountain of storm debris. Some plants show recovery quickly. Others only after weeks or months, so a wait-and-see approach is best before declaring a plant to be a storm casualty.

The ability of native Caribbean plants to recover from hurricane effects is nothing short of astounding; of course, if they could not do so they would not have persisted in a hurricane-plagued environment. In a few years, much of native scrub or a natural garden will recover, and plants established after the storm will be growing. Even in rain forests, as noted previously in the case of Dominica, most trees will be injured but not killed. Where large canopy gaps were created by downed trees or landslides, pioneer species of plants will be well established within a few years. Nonnative invaders may need to be treated. Pinelands can be quite affected as adult pines may be killed by having their trunks fatally twisted and broken and by invading bark and pine beetles afterwards. Replacement pines reseed, although growth may be slow, taking decades to recover. Some sites affected by storms may never recover their pre-storm forest within multiple human lifetimes. A hurricane can cause one plant community to be replaced by another. This is the way of Caribbean nature.

Birds and other wildlife also suffer from hurricanes. The endemism and limited range of many Caribbean species mean that a local disaster can wipe out a population. Perhaps the greatest conservation contribution that can be made post-hurricane is to take care of

Hurricane effects on property and habitat.

the birds, butterflies, and other wildlife, as discussed later in the chapter. Particularly, this is a time when supplemental feeding can be exceptionally beneficial. The sooner vegetative cover, water, and food sources are restored, the better it will be for the native wildlife that depend on them.

4.5 Pest management

Ecosystem restoration and natural gardening are attempts to bring nature back including insects, spiders, and other invertebrates and the birds, lizards, and frogs that eat them. Recovered areas and natural gardens are meant to be insect farms. Pesticides not only kill indigenous, perhaps endemic, insects but in doing so also kill bird and lizard food. Some pesticides can kill or sicken nontarget animals directly or harm people consuming garden produce. Areas of diverse native plants provide ample supplies of natural pest predators. Pesticides are not suitable for use in restorations and natural areas. Nonetheless, the Caribbean tropics can have lots of insects, some of which may be considered pests by people. Natural controls are best.

The tropics are not only full of insects but also, unlike in temperate areas, some sort of insects will be active year-round consuming plants. What should be done about a plant that appears to need help managing its pests? The first consideration is whether a plant needing help is worth keeping. If a species or variety proves repeatedly susceptible to pests or disease, it may be better to replace it with one that is hardier in that particular

Natural insect control, a mangrove cuckoo eating a caterpillar in Grenada.

environment. Native plants and diverse lawns tend not to have overwhelming pest issues. Natural infestations may occasionally strip a native plant of its foliage, but recovery is usually swift. Rare and endangered specimens may deserve special attention if needed. But for native plants in a natural garden or restoration, if one plant dies, another plant can take its place.

Edible vegetables, fruits, and herbs are another matter and may take defending. However, some fruits or vegetables might be shared with birds or iguanas; some may need bugs washed off; fruits might not look store-bought perfect. But this food also is not loaded with pesticide residue. To deal with disease and insects, gardening books have lots of good advice such as using disease-resistant varieties, spacing plants, pruning to allow air movement, planting related species together, crop rotation, removing infected plant parts, soil sterilization, hand picking large pests, netting and caging plants, bagging fruits, encouraging predatory insects, and nematodes. So, there is a lot that can be done before turning to chemicals.

If pesticides are desperately needed in an edible part of the garden, they are applied exactly to the specific pest, not broadly. The best all-round garden help is oil/soap spray applied directly on the affected area, which kills insects by interfering with their breathing. Unlike in temperate zones, vegetable oils can be too heat-producing for use in the tropics. To avoid burning, a mineral-oil-based "horticultural" oil is applied in the cool of the evening. Neem oil kills or repels plant-fluid-feeding insects, such as aphids and their ants, nematodes, and borers. Other natural sprays have insecticidal or repelling properties, such as infusions of hot pepper, garlic, mint, or tobacco. Hot pepper and mint spray can dissuade rats. Copper, iron, and salt treat slugs and snails. Diatomaceous earth kills crawlers such as ants.

Although integrated, natural, limited-chemical pest control in the tropics is usually needed in food production, in natural areas and natural gardens, there is not much reason to poison bird food or desirable insects. This understanding now is widespread in horticulture in the temperate zone. A web search of any U.S. state's agriculture department will provide many natural remedies for home pest control. But in the Caribbean, killing insects has been a way of life. A book as recent as 1985 advocates for Caribbean gardeners to use widely banned DDT, chlordane, and systemic pesticides that enter the plant and then the fruit and then the eater. Any general application of a pesticide kills far more than the target including those that would prey on it. Pesticides need not be organochlorides or organophosphates to be harmful. For example, Bt, an organic-gardening favorite bacterium used to kill target insects, is as much of a pesticide as is any chemical. Different Bt strains affect different but multiple species of flies, caterpillars, or bees. Bt can be useful when selectively applied for a specific, restricted target, such as mosquitoes or tomato hornworms. But this well-accepted treatment is a take-home lesson on pesticides in general. Even Bt can take out beneficial organisms related to those being targeted.

Competition for produce is not limited to insects. In the tropics, black and brown rats can devastate a season of production in a few nights. There seems to be an unlimited supply of rats; likely they first came ashore with Columbus. While poisoning and trapping

European rats and mice, even if cute when young, devastate gardens and harm native wildlife and their habitats.

provide satisfaction, there is no final answer to getting rid of them unless an entire island is treated, which has been done to great effect on small Caribbean seabird-nesting islands. On populated islands, rats are attracted to one yard over the other by food they find there, including fruits and vegetables, compost, trash, or pet food. Control starts with denying them food. One source may be seeds and other food set out for birds. Feeders can be taken in at night. Rats especially show up near harvest time. Thwarting them may mean picking produce just as or before it ripens if it is a sort that ripens off the plant and disposing of unused fruit rather than having it lay about. Smaller plants, sprouts, and seedlings can be protected with wire cages. Rats dislike hot pepper and mint smells, applications of which can be purchased. Rat poisons that work as an anticoagulant are dangerous to dogs and birds that might eat the poison or the poisoned rats. Only rat poisons safe for wildlife and people should be used, within a safety container. As there is no end to available rats, vigilance is constant, and expectations need be reasonable. In some locations, green iguanas and land crabs have similar impacts on yards and gardens. Caging plants often works. Wild hogs, goats, and burros are another matter altogether, requiring stout barriers.

4.6 Specialized gardens

As noted previously, a restoration, public park, or private garden usually needs to accommodate multiple uses. The natural area can be less than the entire lot. In a large enough space, several natural gardens could be established, perhaps emulating different local habitats such as forest, scrub, or xeric. Beyond restoration or natural garden plots, the overall area might be subdivided into subgardens or sections serving different functions and being managed differently. Specialized subgardens might include butterfly gardens, hummingbird gardens, fruit-tree orchards, edible gardens, herb gardens, medicinal gardens, and more. All add habitat diversity.

4.6.1 Butterfly garden

Butterflies are usually considered to be highly desirable wildlife as they entertainingly flit through the air from plant to plant in broad daylight and serve pollinator duty as well. Bees and other pollinator insects follow suit. Bees, in fact, are the most important pollinators in the tropics. Moths, too, are attracted to native plants, although mostly at night. Such plots are often called pollinator gardens. Many pollinator species face conservation issues; so, their special needs might be considered by installing supportive plants in natural plantings or in a butterfly and bee garden of their own. Interest in butterflies is growing. Caribbean-specific information can be found in a few books and on online sites, public gardens, local environmental centers and conservation organizations, and agricultural outreach offices.[33–36] Butterfly attractions, such as those on Aruba, Saint Martin, Saint Thomas U.S. Virgin Islands, Cuba, Martinique, and butterfly gardens at botanic gardens and nature centers provide identification and husbandry information, especially for common local species.

There are 2 varieties of butterfly plants, host and nectar. A nature-friendly planting needs both. Where applicable, these are noted in plant-species accounts in Chapter 5. Butterflies

Two-spotted prepona butterfly.

and moths lay their eggs on host plants where caterpillars mature, eating the plant. Butterflies and moths are quite specific as to their host plants as they have coevolved. Practically speaking, as the host's role in a butterfly garden is to be eaten, they are best installed in the back of a bed. Other plants provide nectar, usually through flowers, attracting pollinators. To produce large, productive crops of flowers, butterfly gardens need to be in the sun. Nectar plants are well placed where their flowers and pollinating action can be observed. Other pollinating-insect enhancements, that also bring them into view, include shallow water sources, patches of damp soil, salt deposits, sugar-water feeders, and insect houses. Supplemental and overripe fruits can be placed in a tray, saucer, and hanging basket or on top of a wall. Bees can be attracted with composite and other bee-friendly flowers, shallow water sources, and various bee shelters. Pesticides have no role near pollinator gardens.

4.6.2 Hummingbird garden

Nearly 20 species of hummingbirds occur in the Caribbean, most with restricted ranges. Many come to butterfly gardens stocked with appropriate host and nectar plants. They sometimes become tourist attractions. Botanical gardens, nature centers, hotels, and tourist spots across the Caribbean have installed hummingbird gardens and put out sugar-water feeders. At some places, visitors can hold a drinking tube for them. These gardens demonstrate how people appreciate hummingbirds and how readily hummingbirds accommodate to people.[14] Hummingbird gardens can be part of a large patch managed to attract birds with other sorts of feeders and water. Such publicized gardens are sought out by birding tourists, much to local economic benefit. Hummingbirds use many of the butterfly nectar plants but not all. Hummingbirds tend to forage at the shrub and tree level so low-stature butterfly flowers are less used. Each hummingbird has a distinctive bill corresponding to flower shapes it can use effectively. Some are generalists but others use only co-adapted flowers, so it is crucial to choose flowers found on a specific island used by local hummingbird species. Important generalized butterfly nectar plants also used by hummingbirds include porterweed, firebush, and swampbush. Hummingbirds also eat pollen, ants, aphids, beetles, gnats, mosquitoes, spiders, wasps, and other small invertebrates, which also can be provided by native plants. Other hummingbird host and nectar plants are noted in Chapter 5. Beyond a dedicated hummingbird garden, hummingbird plants can also be included in a butterfly garden and in a natural subgarden or used in containers.

What about sugar-water feeders? Sugar water does not replace the nutritional complexity of natural nectar, but it does provide energy; and hummingbirds need a continuous supply of high-energy food. Hummingbirds, bananaquits, and other nectar-seeking birds mostly forage from plants; but sugar-water feeders can lure them out to places where they can better be seen. If the area is large enough, separated feeding areas reduce competition among territorial birds and allow resident hummingbirds to create their foraging traplines. Sugar water (1:4 sugar to water ratio) is presented in a hanging tube feeder, a dangling syringe barrel, pan, or bowl. Perches allow non-hovering bananaquits, blackbirds, orioles, and warblers access. Feeders also service bees and other nectar-seekers including bats. Unwanted ants will foul sugar water quickly and so need to be inhibited by ant moats of water blocking the way to the feeder. Sugar water goes bad quickly in the tropical heat, and feeders need to be replenished and cleaned regularly.

4.6.3 Edible garden

Much of the domestically used food grown in the Caribbean comes from private and household gardens, community gardens, and smallholder agriculture traded at market.

Bahama woodstar drinking nectar from a firebush flower.

Dale Browne of Sejah Farm of the Virgin Islands supplies sustainable, locally produced food to residents of Saint Croix.

Most island governments encourage local food production in part to offset the high cost of imported food and the food insecurity and balance of payment issues this presents. Upwards of 80% of the food supply for some Caribbean nations is imported. Information on edible gardening in the Caribbean is available from many sources including national agricultural departments.[9–13] Chapter 8 provides useful references.

Edibles are often best grown in a separate area where their needs can be attended to away from native plants. There likely is no mystery as to what a family might want from its own edible garden. Tomatoes and peppers top most lists as there is nothing like the umami of a homegrown tomato nor is any store-bought scotch bonnet quite as satisfying. Temperate crops can be a challenge. Temperate root crops, adapted to long-daylight growing season and pest-killing winters, prove difficult in the tropics where tubers are attacked by insects and fungi continuously. And why bother as there are so many traditional tropical root crops?

Some traditional edible crops have wildlife value, providing shelter and insects, fruit, and seeds to share. Various beans, especially pigeon peas and long beans, and peppers, especially bird peppers, also are eaten by birds. Vines such as bitter melon, christophine (a squash of many names), and giant granadilla (a passionvine) add structure. Pineapples provide sharp-leaved protected sites. Tropical root crops provide cover. Cassava (manioc, yucca) provides tall, open structure; callaloos (taro/dasheen, tannia/malanga and prickly callaloo) provide leafy intermediate structure; and boniato provides extremely dense, low, viny structure covering the ground. Tropical ground provisions' tubers and leaves naturally have protective chemicals, such as cyanide or calcium oxalate, that for human consumption must be removed from some varieties by processing. But such chemical defenses are what tropical vegetables are all about.

Considerations for edible gardens in the Caribbean include using raised beds; irrigating with drip systems from water storage cisterns, tanks, or rain barrels during morning hours; keeping related plants together and rotating them around year-to-year; removing and

Raised vegetable garden.

burning spent or infected plants; soil solarization between crops; protecting leaves from water-borne soil bacteria and fungi; using locally derived seed, cuttings, and tubers; and amending soil with compost (quite opposite of the recommendation for natural areas). The balancing act of tropical pest control was discussed previously. A Caribbean edible garden need not be tidy. Traditional Caribbean Creole gardens are diverse, crowded, interspersed, and stacked from fruit trees down through edible shrubs to medicinal and edible plants. Raised beds do make things easier.

4.6.4 Herb garden

Caribbean cooking depends on herbs, and growing them at home is nearly expected. As is the case for edible gardens, there is no lack of informational resources.[16–19] In landscaping design, the closer culinary herbs are to the kitchen, the more they will be cared for and used. Common kitchen herbs can be planted together in a dedicated herb garden, interplanted with vegetables, scattered about, decoratively placed in containers, or all of the above. Some typical Caribbean herb-garden plants include basil, chives, cilantro/coriander, chadon beni (culantro), garlic, ginger, mint, rosemary, scallions, thyme, and turmeric. Of course, other culinary herbs are from trees, such as allspice, annatto, bay, cinnamon, and nutmeg and mace, and from native medicinal plants, discussed later.

Herb gardens are wildlife gardens; many herbs are butterfly plants. When left to go to flower, nectar plants include basil, celery, chives, cilantro, dill, fennel, garlic, garlic chives, hyssop, lavender, marjoram, mints, oregano, parsley, rosemary, sage, and thyme. Caterpillar host plants include any of the umbellifers (celery, carrot, cilantro, parsley) for swallowtails, basil for noctuid moths, and mint for hairstreaks and brush-footed butterflies.

Planting extra herbs as hosts for butterflies, perhaps in dispersed places other than the herb garden, supports these species.

4.6.5 Diverse lawn

As previously argued, monoculture sod lawns are challenging in the tropics and unwelcome in a landscape intending to support native plants.[37–39] But admittedly there are places, such as activity areas and pathways, where a lawn is needed. This is the place for a diverse lawn, one composed of multiple species of mostly volunteer, short, mowable plants. A useful way to use diverse lawns is to fill the space between pavers, producing both a lawn and a patio. Diverse lawns also present an opportunity to replace rock, concrete, or asphalt driveways and parking spaces with permeable turfstone lattice having interspersed diverse lawn. The only maintenance for a diverse lawn is infrequent mowing at a high wheel setting. Dry-season watering is unneeded and undesirable, as are fertilizer and pesticides. Having a diversity of plants with asynchronous leafing, flowering, fruiting, seeding, and insect cycles, diverse lawns provide food for bees, birds, butterflies, frogs, lizards, moths, and other wildlife year-round. Lawn plants are always dying for one reason or another; but replacements readily take their place by spreading rhizomes or by self-seeding, recapturing any open ground and keeping the lawn alive, green, and diverse.

4.6.6 Medicinal garden

Also called heritage and folk gardens, medicinal gardens can in fact produce medicines. A medicinal garden provides culturally significant, utilitarian, and mostly native plants

Diverse lawn growing in the spaces between pavers at the National Botanic Garden of Cuba.

grown for their natural products typically used as tonics, teas, or poultices. Traditions come from many sources – Indigenous, European, African, Asian – and have been maintained and passed down through generations. The original European botanic gardens were called physic gardens, meant to supply medicines, some of which came from the Caribbean. Islands and communities have differing cultural traditions and uses for various plants, both native and imported. Important often-used native species include featherbed, greenbriers, granny bush, Guinea hen weed, porterweeds, sages, snakebush, and strongback. Traditions may center around a specific presentation, often multi-species recipes, such as root tonics in Jamaica. In recent years, such folk knowledge is increasingly appreciated, validated, and taught.[18–22]

Mango tree with fruit.

These plants can be gathered into a single medicinal garden. Public and botanical gardens and environmental centers might honor and teach a community's cultural plant heritage through demonstration medicinal or heritage gardens. Medicinal gardens at home likely focus on specific plants the family might use. Sometimes, it is hard to find a native plant that has not been used in traditional medicine somewhere, somehow. Many uses are noted in Chapter 5 and in references in Chapter 8. Most of these plants are native and can be planted in natural garden areas as well.

4.6.7 Orchard

Fruit trees tend to be widely dispersed within the Caribbean landscape as single specimens in yards, along roads, in parks and old homesites, and also interspersed within mostly native habitats. Annona, bananas, citrus, cacao, coffee, papaya, plantains, and sapodilla are found or planted within forests. In a tamer location with enough space, fruit trees can be laid out in a more regular fashion forming an orchard (grove, huerto, verger), which facilitates harvest and care. Neither the design nor maintenance requirement of a fruit grove is appropriate for native portions of a garden. Government programs encourage fruit-tree planting. Common fruit trees include avocado, breadfruit, genip, guava, jackfruit, mango, starfruit, sweetsop, tamarind, and others. Bananas and papaya are tree-sized fruiting plants. There is hardly a private or public garden in the Caribbean without fruit-bearing trees and other tree-sized edible plants. In addition to supplying fruit that animals can share, fruit trees tend to host diverse communities of insects that birds and other wildlife also feed on. In addition to wildlife food, fruit trees can provide shade, shelter, and structure. Native trees and shrubs, such as firebush, cocoplum, and palms, can be interspersed within or allowed to colonize an orchard and their fence rows.

The placement of nonnative fruiting trees should be decided early in planning, as they consume space that constrains natural plantings. They need sun and regular pruning to keep them in their space. Fruit trees might best be along the periphery of a yard, perhaps towards the rear where their produce might not need to be shared with passersby. How to deal with fruit trees existing prior to designing a garden can be a dilemma.

Quite a few native trees and shrubs bear human-edible fruit, noted as edible in Chapter 5, such as cocoplum, guanaberry, hog plum, mastic, satinleaf, seagrape, and wild dilly. Native edible-fruit species can be incorporated into natural areas. As commercial fruit trees require fertilizing, pest management, and perhaps – but not all – supplemental water in the dry season, commercial fruit trees and native trees are best not planted together.

4.6.8 Entrance garden

One part of a garden often overlooked in planning, yet the one that sets its stage, is the entrance and walkway leading to the house. It may be a difficult, well-trodden, perhaps shaded place. It is a good opportunity for containers. Such an area might be designed to reflect the rest of the garden hinting at designs, philosophy, and perhaps lessons from the garden. Native plants can be featured.

4.6.9 Scent garden

Another overlooked potential subgarden is one concentrating on providing scents to the landscape. Scent-bearing plants need not be confined together; but if they are, especially near the entrance or doorways, they may be especially impactful. Public gardens may create scent gardens for the visually impaired. The concept is sometimes expanded to be a sensory garden that incorporates sound, touch, and taste. But for the sighted too, impressions are strongly influenced by scented plants especially flowers including night-blooming species, such as night-blooming cactus. Many herbs are pleasantly scented and so can serve a dual purpose. Native plants that might be considered include braceletwood, citrus of various varieties, fiddlewood, ink wood, Jamaica caper, joewood, marlberry, prickly ash, spicewood, the velvetseeds, West Indian lilac, white torch, willow bustic, and yellow wood. Fragrant plants are noted in Chapter 5.

4.6.10 Xeric garden

Also called xeriscape or arid, cactus, dry, rock, or succulent gardens, xeric gardens are very Caribbean as they can emulate dry and seasonally dry plant communities and provide wildlife values as well. Importantly, they save irrigation water. Their ease of care and aesthetic compatibility vary with the dryness of the site's location – they look natural in dry areas but can be established most anywhere on well-drained sites. Some, such as opuntia cactus with its fruit and pads, are edible. Xeric gardens provide flowers, insects, small fruit, and thorn-protected roost and nesting sites. The dry side of islands, coastal areas, dunes, and other sandy sites support xeric plants, and these are places to search out and find native plants and their seeds for natural dry gardens, so long as it is legal and ecological to do so. Endemic arid-area plants, especially cactus, agaves, and cycads, are among the more threatened Caribbean species. Xeric gardens usually feature nonnative species such as euphorbs and nonnative agave. There really is no need, given the array of sensational cactuses and other native dry-site plants available in the Caribbean.

4.6.11 Container and raised bed gardens

Not all plants need to be in the ground. Some, including native, animal-friendly species, can do just fine in raised gardens, pots, boxes, and planters. Containers can be expensive or free, fancy or simple, pottery or plastic, or a recycled car tire. Any sort of recycled container can be used so long as it drains. Sometimes, the more diverse and creative the containers, the more interesting they are. These can be placed where convenient, collected together, or scattered decoratively. Edible and herb gardens may best be in raised beds,

Turk's cap cactus in a xeric garden in Saint Croix.

Repurposed car tire serving as a flower container garden.

Window-ledge containers with bougainvillea.

where they can be tended. One need not think small. Citrus historically are in containers in European gardens; the authors have a Mexican lime tree that has been in the same pot into its second decade. Raised beds are larger soil- or compost-filled boxes that are useful in overcoming site issues such as surface rock, previously abused soil, or artificial fill. There needs to be a bottom or tree, and shrub roots will aggressively invade the box's good soil. Containers can be used for native plants, with great effect for butterflies, hummingbirds, and other wildlife.

4.7 Wildlife enhancements

Recovering nature is creating naturalistic, native-plant-based habitat for birds and other animals. The animals seek out habitat having their preferred foods and structure. One of the most important restoration lessons is that the food that birds and other wildlife depend on is primarily in the plants, seldom supplemental bird-seed dispensers or cut oranges on a railing. Between a healthy natural garden and supportive nearby environment, birds seldom need supplemental care and feeding. Their food is in the plants. Many people want to also offer enhancements such as feeding stations and water sources so as to bring the birds better into view. Nature centers, botanical gardens, and resorts often provide such opportunities for their visitors.

Birds and other animals that do not mind coming to a feeding station will be attracted by the specific foods each eats – fruits, seeds, nectar, invertebrates, or combinations. Birds attracted to feeding stations are noted in the species accounts in Chapter 6. Perhaps, the most appreciated feeder provides sugar water for hummingbirds from a tube, such as a bottle, tube, or syringe. The best feeding station for seeds and fruit in the tropics is a tray or platform on the ground, attached to the top of a rail or garden wall, on its own pole, or hung, although some birds will come to stationary trays but not mobile hanging trays. Food can also be placed directly on top of fences, on sticks, or on the ground. A seed feeder is a hanging tube or other device that emits seed on demand. Another, mentioned before, is a tube or other reservoir that provides sugar water through a small hole.

Food candidates for tray feeders include banana, orange, mango, guava, fruit jelly, and peanut butter. In trays or in tube feeders, sunflower, millet, rice, peanuts, and various nuts attract different birds. Black oil sunflower seed has weak shells and high-energy content and can be used whole by nearly any species of seedeater as long as it can crack it open. Cracked corn can be thrown on the ground for pigeons and other ground-feeders but becomes toxic in tropical dampness if not consumed right away. Smaller grains (millet, nyjer, and rice) attract smaller seed-eating species such as finches and grassquits and also wintering species. Breaking rice grains into small pieces allows smaller birds their share. Beware of using seed mixes in restricted-access feeders. Birds will toss out the less-desired seeds such as whole corn, milo (sorghum), and wheat. These offerings are best on the ground. A cage around the tube feeder selects for smaller birds. Rats are a persistent problem dealt with best by filling feeders with only what will be consumed in a day and taking them in at night, although mint and pepper sprays have some efficacy.

Tray feeder at Rocklands Bird Sanctuary in Jamaica attracts a Jamaican oriole and orangequits to bananas.

World's smallest bird, the bee hummingbird, taking sugar water from a syringe feeder in Cuba.

Tray feeder with oranges and corn.

Tube seed feeder with red-winged blackbirds, Great Abaco, The Bahamas.

In the Caribbean, supplemental feeding of birds and other wildlife is extremely valuable after disruptive events and essential after hurricanes. Small-island populations may be put at considerable risk to such catastrophic events. Although hurricanes seldom kill off large numbers of terrestrial birds directly, they have indirect impacts; they strip insect-bearing leaves, nectar-supplying flowers, and fruits; they defoliate and shred branches eliminating cover and shelter; and they topple trees. In 2017, a hurricane killed or damaged 30–40 million trees in Puerto Rico, and it is recognized that artificial feeding programs in the aftermath were crucial to sustain affected bird populations. Hurricanes instantly make the habitat unrecognizable, cause floods, and spread salt, putting surviving birds and other animals under stress as they have lost both habitat and food. Some plants leaf-out within a few weeks and flower and fruit within a few months after a storm; some skip the next flowering and fruiting season; some take a few years to get back on schedule; others die slowly. Beyond the immediate aftermath, the months and years after a hurricane continue to have reduced food availability, especially impacting seed-, fruit-, and nectar-eating species. Insect-eating birds have immediate difficulties, but insect populations recover quickly and may flourish after a storm. Predators may find feeding easier post-storm as prey are more visible and vulnerable. A norm in the Caribbean should be to feed birds after hurricanes by putting out sugar water, fruits, and seeds; providing fresh water; and re-installing nest boxes in the months following a storm.

Along with feeding stations, nature lovers like to put up artificial structures to encourage birds to nest. As is the case with food, the primary reason a bird will nest in a garden is because the right plants and habitat structure are present that securely and safely support their nests or provide cavities and appropriate nesting materials. Birds that build nests need cover and security in thick, dense places such as hedgerows; large birds need sturdy branches high in trees. Secure platforms or baskets can be helpful. Nest boxes will be used by species that naturally nest in cavities, usually those made by woodpeckers, and are willing to substitute a nest box. Each species favors certain hole size and box dimensions. Although the knowledge of temperate species can provide guidance, guessing and experimentation may be needed for most Caribbean cavity-nesters.

Nest boxes have proven essential in conserving cavity-nesting species threatened by loss of trees including Caribbean trogons and parrots. Through nest boxes and captive rearing, Puerto Rican Parrots, the population of which had declined to fewer than 2 dozen in the wild in the late 1960s, has risen to the low hundreds in the wild and captive populations, despite habitat-devastating 2017 hurricanes.[40] Woodpeckers, parrots, parakeets, trogons, some owls, and some flycatchers are among other typical nest box users, as noted in species accounts in Chapter 6. Some bats species roost and reproduce in properly designed boxes. Finding out what bats are about and how they roost might suggest what, if any, house design they might use.

Houses are a supplement; aways natural cavities and cavity plants need to be protected, especially single-stemmed palms as well as trees with branch knots, dead branches, and decaying trunks. If at all possible, considering human and structural safety, dead tree and palm trunks should be left standing as they supply both cavity space and food. Boxes are more than just for birds and bats. Even if birds shun a box, bees, frogs, lizards, and other animals can move in. They are wildlife boxes.

A nest box for birds or more broadly for various small animals.

One of the more important benefits a garden can provide to birds, butterflies, and other desirable wildlife is making water available. The drier the location, season, or year, the more important a water source will be to animals. On some islands, available water sources dry up in the dry season, and on others, availability decreases markedly both in amount and in distribution. The most useful water sources are rain gardens and small ponds having aquatic vegetation and surrounded by the cover of other plants. Smaller water features, such as birdbaths, are also best on or in the ground and near or under plant cover. Nearly any shallow vessel can be a water source for various animals, including small basins, ceramic dishes, livestock bowls and troughs, or small shallow pools – generally, the shallower the better, taking the leg length of the expected users into account. Muddy spots and shallow pans with exposed rocks can be used by butterflies and other insects. Birds seem to like a bit of moving water; a faucet drip will do although solar-powered aerators are also available. Damp spots and small ponds are surprisingly easy to make; all it takes is a thick plastic sheet lining a depression and a bit of water.

Mosquitoes can be an issue for standing water; zika, chikungunya, and dengue are not to be trifled with. One approach is to empty and refill a small water feature every 4–5 days, before the eggs or larvae have the time to mature. The biological control Bti (spores of

Bacillus thuringiensis israelensis) is rather specific for mosquitoes and some other flies and can be used in most water sources. For large features, fish can be added to eat mosquito eggs and larvae, the best being topminnows (*Gambusia*), also rightly called mosquitofish. Species are native to The Bahamas and Greater Antilles, and related North American mosquitofishes (*Gambusia affinis, Gambusia holbrooki*) have been introduced in the Caribbean precisely for mosquito control. Garden ponds need to be escape-proof for fish outside their native range.

References

1. Gann, G. D., T. McDonald, B. Walder, J. Aronson, C. R. Nelson, J. Jonson, J. G. Hallett, et al. 2019. "International Principles and Standards for the Practice of Ecological Restoration. Second Edition." *Restoration Ecology* S1–46. DOI:10.1111/rec.13035
2. Perrow, M. R., and A. J. Davy, eds. 1999. *Handbook of Ecological Restoration. Volume 1. Principles of Restoration*. Cambridge: Cambridge University Press.
3. Higgs, E. 2003. *Nature by Design: People, Natural Processes, and Ecological Restoration*. Cambridge: MIT Press.
4. Canadian Parks Council. 2008. *Principles and Guidelines for Ecological Restoration in Canada's Protected Natural Areas*. Parks Canada and the Canadian Parks Council. parks.canada.ca/nature/science/conservation/ie-ei/re-er/pag-pel
5. Cairns, S., N. Dudley, C. Hall, K. Keeneleysie, and S. Stolton. 2012. *Ecological Restoration for Protected Areas: Principles, Guidelines and Best Practices*. IUCN. www.iucn.org/resources/publication/ecological-restoration-protected-areas-principles-guidelines-and-best
6. WWF International. 2003. "Global Partnership on Forest Landscape Restoration: Investing in People and Nature." IUCN, Forestry Commission of Great Britain. www.forestlandscaperestoration.org/
7. Locatelli, B., C. P. Catterall, P. Imbach, C. Kumar, R. Lasco, E. Marín-Spiotta, B. Mercer, J. S. Powers, N. Schwartz, and M. Uriarte. 2015. "Tropical Reforestation and Climate Change: Beyond Carbon." *Restoration Ecology* 23:337–43. DOI:10.1111/rec.12209
8. Pancel, L., and M. Köhl, eds. 2016. *Tropical Forestry Handbook*, 2nd edition. Berlin: Springer.
9. Attfield, H. H. D. 2001. *A Tropical Guide to Year Round Vegetable Gardening*. Arlington: Vita Technical Publication.
10. Ferguson Fowler, C. H. 2013. *Backyard Gardening in the Bahamas and the Caribbean*. Published by Author.
11. Hall, S. 2011. *The New Caribbean Home Garden Handbook*. London: Guardian Media, Ltd.
12. Winters, H. F., and G. W. Miskimen. 1967. *Vegetable Gardening in the Caribbean Area*. US Department of Agriculture, Agriculture Handbook 323. Washington: Department of Agriculture. www.naldc.nal.usda.gov/download/CAT87209120/PDF
13. Ford, J. R. D., D. Ramroop, V. Boerger, V. Lopez, A. Pantoja, J. James, and O. Elabanjo. 2013. *A Vegetable Garden for All*, 5th edition. Antigua and Barbuda: Zero Hunger Challenge. www.fao.org/docrep/019/i3556e/i3556e.pdf
14. Green SXM. 2019. "Planting a Caribbean Wildflower and Pollinator Garden." greensxm.com/caribbean-wildflower-and-pollinator-gardens/
15. Reese-Warren, M. 2021. *The Ecological Gardener: How to Create Beauty and Biodiversity from the Soil Up*. White River Junction: Chelsea Green Publishing.
16. Belsinger, S., and A. O. Tucker. 2019. *Grow Your Own Herbs*. Portland: Timber Press.
17. Creasy, R. 2019. *The Kitchen Herb Garden: Growing and Preparing Essential Herbs*. Clarendon: Tuttle Publishing.

18. Mitchell, S. A. 2011. "The Jamaican Root Tonics: A Botanical Reference." *Focus on Alternative and Complementary Therapies* 16: 271–80.
19. Clement, Y., P. Sutherland, P. Moodley, and B. Chevannes. 2014. *Herbal Medicine Practices in the Caribbean*. New York: Routledge.
20. McCormack, J. H., K. Maier, and P. B. Wallens. 2011. *Bush Medicine of the Bahamas*. Charlottesville: JHM Designs Publications.
21. Wilmanowicz, R. 2008–2010. *Bush Medicine in the Bahamas: A Modern Approach*. Germany: Create Space.
22. Mahabir, K. 1991. *Medical and Edible Plants Used by East Indians of Trinidad and Tobago*. San Juan Trinidad: Chakra Publishing House.
23. Degras, L. 2005. *The Creole Garden: Cultural, Scientific and Technical Landmarks*. Pointe-a-Pitre: Jasor Editions.
24. Kushlan, J. A., and K. Hines. 2014. *Attracting Birds to South Florida Gardens*. Gainesville: University Press of Florida.
25. Buchanan, R. 2000. *Taylor's Master Guide to Landscaping*. New York: Mifflin Harcourt.
26. Holden, R., and J. Liversedge. 2014. *Landscape Architecture: An Introduction*. London: Laurence King Publishing.
27. Uncapher, A., and C. Woelfle-Erskine. 2012. *Creating Rain Gardens: Capturing the Rain for Your Own Water-Efficient Garden*. Portland: Timber Press.
28. Steiner, L. M. 2012. *Rain Gardens: Sustainable Landscaping for a Beautiful Yard and a Healthy World*. Beverly: Voyageur Press.
29. Wallace, T. 2008. *The Rain Garden Planner: Seven Steps to Conserving and Making Water in the Garden*. Atglen: Schiffer Pub Ltd.
30. Bainbridge, D. A. 2015. *Gardening with Less Water*. North Adams: Storey Publishing.
31. UNEP. 1997. *Source Book of Alternative Technologies for Freshwater Augmentation in Latin America and the Caribbean*. Washington: Unit of Sustainable Development and Environment General Secretariat, Organization of American States.
32. Taylor, B. N., E. Stedman, S. J. Van Bloem, S. L. Whitmire, and S. J. DeWalt. 2023. "Widespread Stem Snapping but Limited Mortality Caused by a Category 5 Hurricane on the Caribbean Island of Dominica." *Forest Ecology and Management* 532. DOI:10.1016/j.foreco.2023.120833
33. Riley, N. D. 1975. *A Field Guide to the Butterflies of the West Indies*. New York: Quadrangle Field Guide Series.
34. BOC. 2023. "Butterflies of Cuba." www.butterfliesofcuba.com
35. Garraway, E., and A. Bailey. 2005. *Butterflies of Jamaica*. London: Macmillan Caribbean Natural History.
36. Turner, T., and V. Turland. 2017. *Discovering Jamaican Butterflies and Their Relationships Around the Caribbean*. Safety Harbor: Caribbean Wildlife Publications.
37. Daniels, S. 1995. *The Wild Lawn Handbook: Alternatives to the Traditional Front Lawn*. New York: Macmillan Publishing Company.
38. Penick, P. 2013. *Lawn Gone! Low-Maintenance, Sustainable, Attractive Alternatives for Your Yard*. Berkeley: Ten Speed Press.
39. Wormser, O. 2022. *Lawns Into Meadows, 2nd Edition: Growing a Regenerative Landscape*. San Francisco: Stone Pier Press.
40. Koches, J. 2022. "A New Step Forward in the Future Recovery of the Puerto Rican Parrot." www.fws.gov/press-release/2022-03/new-step-forward-future-recovery-puerto-rican-parrot

5 Nature-friendly plants

5.1 Plants and recovering nature

The multiple roles of plants, mostly native plants, in recovering Caribbean nature, including supporting birds and other wildlife, have been persistent themes through this book. Given the Caribbean's diverse flora, it is easy to be overwhelmed by the wealth of options and underwhelmed by available advice on native plants to use in nature recovery. Plants and groups of plants suggested in this book were chosen by taking into account their potential availability, geographic distribution, and suitability for restorations, and especially their environmental values in recovering nature and supporting birds and other wildlife. Plant availability may limit species choice but what is more important is to use what animal-friendly native plants are available, even if these will create a stand that is different than the original intact plant community.

5.1.1 Availability

As discussed previously, the availability of native plant stock is certainly an issue in the Caribbean where plant nurseries, garden centers, and landscapers still focus on non-native plants because that is what customers have wanted – which would change if the demand for native plants increases. Plant-species accounts note if a plant might be available through plant nurseries, but this is only a suggestion and cannot be counted upon for any particular plant or plant nursery. Potential customers should ask for them. Caribbean plant species also are available for purchase online and through nurseries in Florida and elsewhere, but genetics is an issue with plants foreign to an island. Florida is perhaps the most reliable source for plants native to the Caribbean, as South Florida's indigenous tropical plants originated there. The plants that made it to Florida's tropical hammocks tend to be the Caribbean's weedy, supertramp, lowland and coastal species that also have dispersed and continue to disperse widely within the region. It is unlikely that these vagrant species have genetically diverged much among populations. This is certainly not the case for endemic species and their relatives. It might threaten local genetics to secure a species from Florida that has close and perhaps endemic relatives on an island. To overcome genetic issues, propagation from local plants is a desirable, viable, safer alternative to purchase, and accounts note how a plant might best be propagated.

DOI: 10.1201/9781003396321-6

5.1.2 Distribution

Geographic distributions of many plants recommended for consideration tend to range over large portions of the region rather than being confined to a single island. This is to provide options that are more-or-less region-wide. Focusing on widespread plants inevitably ignores thousands of rare, endangered, or endemic species because their limited ranges lack region-wide applicability. However, many of the plants or plant groups discussed also have endemic species in the region, which will have similar wildlife values as their more widespread relatives. Local sources can offer information on island-indigenous and endemic plants and these should be emphasized, supplementing regional suggestions.

5.1.3 Suitability

Suitability to restoration programs and survival in park or garden settings were considerations in recommending a plant. Many highlighted plants are pioneer species, plants that colonize forest openings and disturbed sites, rather than species found in stable, mature forests or in specialized situations. Plants of shores, scrub forest, xeric areas, and disturbed sites are especially hardy and so can do well when planted. Protected area ecosystem restoration might use pioneer starter species to create habitat rapidly. Or, they might use mature-forest species to give them a head start, bypassing the pioneer successional stage. Hardy native plants are highly recommended for smaller-scale restorations such as private gardens. Tended gardens also present opportunities to include less-flexible, local, endemic species for their conservation value. That many of the plants to be considered are common, widely distributed, adaptable, and good for birds is not a coincidence; they are widely distributed precisely because they have fruits and seeds adapted to being eaten and dispersed widely by birds and bats.

5.1.4 Environmental values

Plants' wildlife values include their ability to support birds, butterflies, and other animals through their structure, flowers, fruits, and insects they provide. A plant's structure – size, shape, leaf and branching architecture, canopy density, rooting system, epiphytes, thorniness, and such – contributes to its being able to provide shelter and cover for birds, lizards, and other animals. Other aspects of a plant's structure include trunk, bark, dead and living branches, cavities, epiphytes, leaves, buds, flowers, and fruit all of which may provide substrate, food, and shelter for invertebrates. These include a myriad species of ants, aphids, bees, beetles, bugs, flies, grasshoppers, snails, spiders, termites, wasps, and, in some situations, worms, millipedes, and other detritus-processing animals. Some plant and animal species have coevolved with their invertebrate commensals. Butterfly caterpillars are a well-known example; specific butterfly species use specific plants or groups of plants as "hosts" for their caterpillars that eat the plants, which have their own chemical defenses that may deter species not adapted to them. Flowers may have structure specific to their targeted pollinators. Ants may specialize in using plant and honeydew-producing insect species symbiotically.

These small animals attract an array of carnivorous animals from spiders and mantis to birds and lizards. Bird species search for small invertebrates differently and so may tend to use different plant species or different parts of a plant. For simplicity, these are referred to as insectivores or insect-seeking birds and other animals. The list of Caribbean insectivorous animals is long, although few might be at any site. To avoid repetition in the species accounts, the insectivore list for a plant might include bananaquits, bats, beetles, catbirds, cuckoos, hummingbirds, flycatchers and kingbirds, frogs, geckos, ground lizards, gnatcatchers, kestrels, lizard-cuckoos, mantis, mockingbirds, orioles, small owls, spiders, thrashers, thrushes, tree lizards, warblers, wasps, woodpeckers, and vireos. More local insectivorous species also might frequent plants; and other species will hunt insects near and over natural areas and gardens, including nighthawks, swallows, and swifts.

Flowers are important sources of animal food. Plants have various and sometimes complicated pollination strategies, some of which rely on attracting nectar-feeding animals, also referred to as nectar-seeking, nectivores, and nectivorous. Nectar-feeding insects in turn provide some of the invertebrate food for insectivorous animals. Some flowers are highly specialized in attracting and managing specific pollinators; others are more generalized, attracting multiple nectivorous animals. The species accounts note plants that attract nectar-seeking animals. Differing with the plant species and location, Caribbean nectar-feeding animals include ants, bananaquits, some bats, bees, blackbirds, butterflies, hummingbirds, flies, moths, orioles, tanagers, vireos, some warblers, and wasps. However much appreciated hummingbirds and butterflies are, bees are the most important pollinator of many tropical plants; and their conservation is a matter of great concern. Honeybees, although nonnative, also are important pollinators of many sorts of plants, which benefit when apiaries are included in natural and other gardens. In addition to nectar, flowers produce pollen, which is collected by bees and eaten by some birds, such as hummingbirds. A few produce pollinator-attracting oils. Flowers and flower buds also are eaten directly by iguanas, parrots, and other species.

Plants have differing seed dispersal strategies. Many Caribbean species involve attracting fruit-eating (frugivorous) animals that disperse their seeds. Caribbean islands have fewer insectivorous birds than might be expected in a similarly sized patch of South or Central American mainland. Birds that on a continent might be narrowly insectivorous, in the Caribbean may be fruit-eating or include fruits in their diet. These include such species as flycatchers, kingbirds, and wintering swallows and warblers. Which species of fruit-eating birds are attracted to a plant depends on such factors as a fruit's size and hardness, fruit-handling capabilities of the bird, season of availability, and overlapping ranges of plants and birds. The fruit of some plants are small and hard, usually called seeds although some are really seed-containing structures. Granivores are attracted to plants such as grasses, legumes, composite flowers and to those with hard capsule-like fruit, and also to seed feeders that homeowners or nature centers might put out for them. Caribbean fruit-eating animals include bananaquits, some bats, blackbirds, bullfinches, catbirds, some flycatchers and kingbirds, gnatcatchers, grackles, iguanas, mockingbirds, orangequits, orioles, palmchats, palm-crows, pigeons, parakeets, parrots, spindalis, some swallows, tanagers, thrashers, thrushes, trogons, some wintering and resident warblers, some woodpeckers, and vireos. Caribbean seed-eating animals include bullfinches, buntings, finches, grassquits, orangequits, pigeons, parakeets, parrots, and sparrows.

5.1.5 Presentation

The species accounts that follow provide information on plants for consideration in programs to recover Caribbean nature and a few on a do-not-plant list. They are listed following currently accepted taxonomic order.[1, 2] Accounts are compact and indicative rather than comprehensive; additional information can be gleaned from online and print resources, some discussed in Chapter 8, and from local information sources.[3–26] The index can be used to find specific plants directly.

Scientific names (genus and species) are provided and sometimes higher taxonomic groupings such as genera or families depending on the plants and plant groups being presented. Scientific names, perhaps surprisingly, are not particularly stable these days as modern genomic and taxonomic studies result in reclassifications and renaming. It takes a while for these new names to be incorporated by the scientific community, longer to get into books, and even longer, if ever, onto plant nursery tags. Because several scientific names may be encountered for the same plant, the account may include additional alternative genus or species names, which may not be those accepted scientifically at the moment.

Common names are provided in English, also in Spanish, French, and Haitian where appropriate. Common names actually tend to be more stable than scientific names although they lack universality, being different among languages and islands and sometimes within an island. Common names provided are not exhaustive for the entire region but are indicative. Local common names reveal much about a plant. Rose in the name means it has a pretty flower. Wood, bark, and root acknowledge the plant parts that traditionally were useful. Torch means its wood carries fire brightly and slowly so branches can provide light when on fire. A portable light is still called a torch in British English. Thatch means that leaves were used for thatching or weaving. Obeah, granny, or the name of an ailment is how a plant was or is used medicinally. Obeah deals with spirits. Tree in the Caribbean has a flexible meaning; sometimes, it may be any plant barely over a meter or so tall. Plants were used differently on different islands and so bear different names. Different plants also were used similarly or have other similar properties and so may share a common name.

5.2 A Do-not-plant list

Before discussing environmentally friendly plants for nature recovery, it is worthwhile to note species not to be planted or to be removed if already there. These nonnative species are invasive, spread easily, and are proven aggressive competitors of native species and disruptors of natural habitats and cultivated areas. The do-not-plant list includes familiar plants; they are familiar because they are weedy and so widespread. Some are well-liked because they are useful, pretty, or part of everyday life. Most were introduced with good intentions to serve some purpose; now, we know their downsides better. They can be replaced with environmentally valuable plants.

Tradescantias (*Tradescantia*): A New World genus with approximately 90 species of dayflowers. Three nonnatives are common: wandering Jew (*Tradescantia zebrina*) originally from Mexico to Colombia, oysterplant (*Tradescantia spathacea*) from Mexico and Central America, and purple heart (*Tradescantia pallida/purpurea*) also from Mexico and Central America. They are widely planted as low ground cover as they have interesting leaves and require no care. Their flowers and fruit attract birds, which spread the seeds allowing them to escape cultivation and invade forest edges, roadsides, savannahs, and coastal scrub. They may be epiphytes or form dense monospecific ground-smothering stands, displacing short-statured native plants.

Marabú (*Dichrostachys cinerea*) (acacia Saint Domingue, bell mimosa, Chinese lantern tree, el marabú, marabu tree, marabou thorn, marabou weed, sicklebush): Originally from the Old World tropics, this is a bush or a small tree up to 7 m tall. It has showy, honey-bee attracting, fuzzy, pink flowers; its edible protein-rich pods can be important cattle forage; and its hard wood can be used as slow-burning firewood valuable especially in locations where other wood is scarce. It has an array of medicinal uses in the Old World. Clearly, it is a tree of many values, at least in its native range. However, it was introduced to the Caribbean in the 1800s and since has proven hard to be controlled as it is fire-resistant and fast-spreading, creating marabou woodlands. In Cuba, it has taken over tens of thousands of hectares of previously agricultural land and roadsides. It remains a highly invasive threat to agricultural lands and seasonal scrub forest throughout the region.

Mesquite (*Prosopis juliflora*) (algarroba, baron, bayahonda blanca, bayahonde, bayawon, cambrón, cashaw, many other names where introduced around the world): This likely is a Caribbean native brought by Indigenous Peoples as food. It is an impressively thorny, small tree with a height of about 4 m that prospers not only in dry environments but also in pastures, coastal habitat, roadsides, vacant lots, native forest, and scrub. It has many uses for live fences, poles, lumber, furniture, firewood, charcoal, animal feed; pods are edible. It also has many wildlife qualities such as a thorny interior that provides well-protected nest sites and trunks and leaves that support insects, particularly grasshoppers. Tiny flowers on spikes attract bees and also butterflies and wasps. Birds, especially pigeons, eat the sweet pods, which is also why it spreads so widely. Although it is native to the region, it is one of the world's worst invasive species that has taken over large areas in Africa, Asia, and North America in part due to its use in antidesertification programs. It is too dangerous to plant purposely in Caribbean restorations.

Orchid trees (*Bauhinia*): These are attractive, well-appreciated leguminous plants. Their common names refer to their orchid-like flower or hoof-like leaves. Poor man's orchid (*Bauhinia variegata*) and orchid tree (*Bauhinia purpurea*) are widespread on some islands along roadways and in parks. Upon ripening, seeds shoot out of the pod, spreading into native habitat. However, they are excellent hummingbird trees. The replacement is the Hong Kong orchid tree (*Bauhinia* x *blakeana*), which is sterile and so does not pose an environmental threat. It grows fast, up to 12 m, with a wide canopy that withstands pruning and shaping. It is propagated by cuttings or air layering and is available from plant nurseries. However, it is essential to be certain of identity so as not to use invasive species.

Velvet bean (*Mucuna pruriens*): Originally from the Old World tropics and subtropics, it is used around the world as a leguminous protein source and also in folk and herbal-based medicine. It is a fast-growing, spreading, annual vine that takes over sites, crowding out other plants. Most forms have fearsome, itch-causing seed pods.

Woman's tongue (*Albizia lebbeck*): Originally from India to South East Asia, this is a tree that seemingly will grow anywhere from native forest to a crack in the pavement. It has long been in the Caribbean but is quite invasive, able to colonize any empty space. It provides little wildlife value, and its rattling pods are noisy.

Australian pine (*Casuarina litorea/glauca, Casuarina equisetifolia*) (beachoak, casuarina, sheoak): Otherwise enchanting trees, they tower above native species along beach dunes, line fence rows, create hedges, or stand as neighborhood landmarks. Actually, they are a scourge of the Caribbean shore. Tiny seeds inside a painfully spiky cone-like fruit are spread by water and wind and so take over shorelines and cleared land. People, growing up in their beachside shade, may consider them with fondness as part of personal history and enjoy their beach shade. But their shallow roots cause coastal erosion, keep sea turtles from nesting, and come down in storms taking the dune with them. The dense litter of needle-like branchlets inhibits native shoreline plants. They offer little environmental value beyond offering tall perch sites for birds. Buttonwood, palms, maho, and seagrape are native beach shade replacements.

Castor bean (*Ricinus communis*): This large semi-herbaceous euphorb was introduced from tropical Africa. It volunteers on forest margins, waste and cleared sites, and roadsides,

Australian pine on the shore in The Bahamas.

often quite obviously. Its seeds contain highly toxic ricin. It provides no value to birds or other wildlife.

Paperbark tree (*Melaleuca quinquenervia*) (broad-leaved paperbark, cajuput, melaleuca, punk tree): This eucalyptus relative was originally from Australia, New Guinea, and New Caledonia, one of many related species from the region and important wildlife trees there. Paperbark is a tree of highly seasonal wetlands. Although its thick pealing bark is intriguing and its flowers are used by bees, it spreads widely through long-lived, tiny, wind-dispersed seeds. Locally, it forms wildlife-avoiding, thick, monospecific stands and particularly poses a threat to Caribbean wetlands, as is happening in the Zapata Swamp in Cuba.

Brazilian pepper (*Schinus terebinthifolius*) (Christmasberry, Florida holly): This destructive originally South American plant invades native forests and can totally take over cleared areas. A poison ivy relative, its leaves, stems, and berries can cause dermatitis and respiratory issues. Birds are attracted to the fruit and spread seeds widely. Once established, it is extremely difficult to eradicate. It is appreciated for its holly-like appearance of red berries offset by green leaves. Brazilian pepper wreaths, and their distinctive pinene scent, can be a Christmastime tradition. But it still needs to be grubbed out at every opportunity and replaced with native shrubs and trees. Pulling seedlings is easiest.

Neem (*Azadirachta indica*) (Indian lilac, lilac tree, margosa): This is a fast-growing, drought-resistant, Southeastern Asian tree that can achieve 20 m in height. It is a mahogany relative used for multiple purposes including shade, food, wood, and medicine for a variety of issues from inflammation to infections. Its fruit and seeds are the source of horticulturally useful neem oil, mentioned in Chapter 4. Its fragrant flowers attract insects, and its fruit attracts birds. It would seem to be a good wildlife tree; but it is widely invasive in forests, disturbed areas, pastures, and roadsides, crowding out native species.

Brazilian jasmine (*Jasminum fluminense*) (Gold Coast jasmine, jasmine, jazmín de trapo, wild jasminum): This African olive relative readily escapes cultivation. Birds eat the fruit-dispersing seeds, and so it is an aggressive invader of forests, gardens, pastures, and open space. It is an exceptionally efficient vine, climbing high into forest canopy and covering host plants, perhaps entire stands of native vegetation.

Creeping foxglove (*Asystasia gangetica*) (Chinese violet, coromandel, Ganges primrose, Philippine violet): Originally from the Indian Subcontinent and Australia, this is a short-statured plant or even a ground cover. It was introduced for its ornamental purple flowers, which do attract nectar-seeking insects. However, it invades nearly everywhere, smothering other plants.

Umbrella tree (*Schefflera actinophylla/Heptapleurum actinophyllum*): Originally from New Guinea and Australia, umbrella trees are often used as a houseplant. In the wild, birds disperse this ginseng relative's small red fruit and seeds. Although it can become a tree with a height up to 18 m, it can start epiphytically on other trees, roofs, and walls, destroying them while sending roots out to destroy pavement and septic tanks. Umbrella trees should not be let out of their indoor pot.

Hawaiian sea grape (*Scaevola taccada*) (Asian scaevola, naupaka, sea lettuce, white inkberry): Originally a coastal shrub of the Old World tropics, in the Caribbean it has been widely planted as a hedge. But it readily escapes cultivation. It outcompetes native dune vegetation including its native relative, inkberry. White fruit, although eaten by birds, are

Hawaiian sea grape.

adapted to float, allowing it to invade new beaches. Buttonwood, cocoplum, and seagrape can be used to replace it as coastal hedges.

White top (*Parthenium hysterophorus*) (famine weed, Santa-Maria, Santa Maria feverfew, whitetop weed): This is a New World daisy-like composite originally from Mexico through South America. However, it has been introduced widely in Australia, Asia, and Africa. It is economically detrimental being highly invasive on disturbed land, roadsides, pastures, and agricultural fields but is toxic to wildlife, cattle, and humans. It is classified as a global invasive species and illegal to sell in European Union nations.

5.3 Plants to consider

Although tightly curated, the species and groups of plants offered for consideration remain rather encyclopedic. For a quick look, Table 5.1 provides a short list of Caribbean wildlife plants, especially suitable for restorations and natural gardens.

Ferns (Pteridophyta): The Caribbean is a center of fern diversity. With over 600 species of ferns and allies, Cuba has one of the world's richest national fern floras.[26] Even much smaller Guadeloupe has 300 species, and Reach Falls in eastern Jamaica, alone, has 45. Ferns are diverse; they may be tree-size or tiny, aquatic or terrestrial, vines or epiphytes. Terrestrial species provide cover and structure along habitat edges and within shaded understories where other plant species are shaded out. The primary wildlife value of ferns is that they provide structure and cover. The largest are the tree ferns. The Caribbean tree fern (*Cyathea arborea*) can reach 9 m in height. It is a hardy, forest-understory plant that can rapidly colonize open sites. It can do well in plantings, providing complex structure at its crown, large leaves, and rough "trunk." Brake ferns (*Pteris*) are short-statured plants

Table 5.1 A short list of Caribbean wildlife plants suitable for restorations and natural gardens

Bay cedar	Myrsine
Blacktorch	Passionvines
Blolly	Pigeon plum
Bloodberry	Porterweeds
Bougainvillea*	Prickly pear cactus
Butterfly bush	Royal palm
Buttonwood	Sages
Coconut palm*	Satinleaf
Cocoplum	Seagrape
Crabwood	Shortleaf fig
Darling plum	Spanish needle
Firebush	Stoppers
Gumbo limbo	Swampbush
Heliconias	Thatch palm
Hog cabbage palm	Tremas
Hong Kong orchid tree*	Velvetseeds
Indigoberry	West Indian lilac
Jamaica caper	Wild coffees
Lantanas	Wild dilly
Lignum vitae	Wild petunias
Marlberry	Wild tamarind
Milkweed	Willow bustic

Note: * denotes nonnative plants.

Caribbean tree fern in Jamaica.

sometimes used as house plants; so, there are horticultural varieties available. They rather typically colonize limestone and stone walls, adding structure there. Maidenhair ferns (*Adiantum*) are attractive, common, and diverse – Jamaica has 23 species, and Puerto Rico 13. They volunteer at water features where they provide cover along edges. Native swordferns (*Nephrolepis*) can be thick edge or sparse understory plants providing low structure and cover. The ubiquitous pantropical swordfern (*Nephrolepis biserrata*) is also a Caribbean native. It is readily available from plant nurseries. The related Asian swordfern (*Nephrolepis brownii*) is also a popular ornamental plant, available from plant nurseries, but it is naturalized and invasive in the American tropics and needs to be avoided. Polypody ferns (*Polypodium*) are diverse in the region. Jamaica has 33 species; Puerto Rico, 27. They are terrestrial and epiphytic, adding structure to trees. The high level of endemism among Caribbean ferns provides opportunities to use local populations in restorations and garden plantings. However, species and cultivars are also available in plant nurseries. Ferns spread naturally by rhizomes and by wind-borne spores, which can travel long distances, which is why some species are widespread, and will volunteer in plantings. They are propagated by division.

Pines (*Pinus*): Caribbean pine forests support unique arrays of plants, birds, lizards, and invertebrates, many of which are endemic and restricted to pine habitat. Restoration, protection, sustainable management, and plant-species conservation in Caribbean pine forests are high priorities. Pines host diverse communities of insects, spiders, and other invertebrates in their needle clusters, cones, and bark as well as in the shrubs, short-statured plants, and pine litter beneath the trees. These are targets of insectivorous birds, some of which are pine specialists such as the Bahama nuthatch and Bahama and pine warblers. Common species include bananaquits, flycatchers and kingbirds, grassquit, tanagers, warblers, and woodpeckers. Bullfinches, parakeets, and woodpeckers extract and eat

Caribbean pine with a Cuban green woodpecker.

pine seeds. Pine trunks are excavated for nesting cavities by woodpeckers and then used by many other species; dead standing trunks are especially valuable for nesting cavities. The trees also provide high perches and nest sites for large birds. Pine plantations have fewer birds and other wildlife than do native pine forests because of their lack of diversity. The predominant species is Caribbean pine (*Pinus caribaea*), which is a limestone specialist occurring broadly from seasonally wet savannas in The Bahamas to premontane hills in western Cuba. Long harvested in the Caribbean, it is an important commercial species as it is fast-growing and so has become the most widely planted tropical conifer worldwide. Cuban pine (*Pinus cubensis*) from eastern Cuba and Hispaniolan pine (*Pinus occidentalis*) are closely related. As a commercial species, Caribbean pine seedlings and saplings are available, although the genetics is not likely to be local. Pines can be used in parks and gardens, best as part of a diverse native garden.

Cedar (*Juniperus barbadensis*) (Barbados cedar, juniper, red cedar, West Indian juniper): Native to The Bahamas, Cuba, Hispaniola, Jamaica, and the Lesser Antilles, cedar is an attractive native evergreen. There are 2 varieties, one now occurs only on Saint Lucia as it is extinct on its namesake Barbados. Both habitat restorations and gardens can use cedar for its wildlife and conservation values. A true cedar, it holds its cones until late into the dry season when other fruits are unavailable to birds. Usually ranging from a shrub to small tree, cedar can also become a tree up to 10–17 m tall. Its dense branches are covered with scale-like leaves that provide exceptional wildlife cover. It has 2 crops of small, 1-mm, blue-green cones annually that attract fruit-eating and seed-eating birds, especially such species as spindalis and tanagers. Fruit may persist through hurricanes. Culturally important, it once was heavily lumbered and is now protected in some jurisdictions. Cone-tea was used for treating fevers and pain and as a mild diuretic. Both habitat restorations and gardens can use cedar for its wildlife and conservation values. It is propagated by softened seed and by cuttings.

Cedar.

Cinnamon (*Canella winterana/alba*) (camille, canela, canella, cinnamon bark – to distinguish it from true cinnamon, pepper cinnamon, poivree, kanèl, white cinnamon, whitewood bark, wild cinnamon, winter's bark): Occurring Caribbean-wide, it was an early commercial discovery brought to Europe by Columbus as a potential substitute for cinnamon, which was highly valued at the time. As a result, it was once heavily harvested for its cinnamon–clove-smelling bark. It is fast-growing to 5 m, and up to 18 m in forests. It is an attractive tree having a thick canopy of dark, shiny leaves, which provides shelter and nest sites. Orange, red, or purple flowers with yellow centers in terminal clusters attract nectivorous animals. The soft, red to yellow, 1-cm winter-spring fruit on female plants ripens singly attracting fruit-eating birds. The tree has been used similar to cinnamon as a flavoring, insect repellent, for indigestion, and as a hard wood. It is propagated by seed and cuttings; both need to be grown in shade.

Annonas (*Annona*): These include common commercial and yard fruit trees that are also superior wildlife trees. They support large, diverse communities of insects and spiders, especially if festooned with epiphytic bromeliads and orchids. The fruit, once ripe and opened, can be eaten in bits by any fruit-eating bird. Pond apple (*Annona glabra*) (anona de manglar, anón liso, cayuda coeur de boeuf, kowosol mawon, mammier) is a Caribbean-wide species naturally occurring around lakes, ponds, and wet solution holes. Its roots have air spaces and so were used for floats. Its large fruit is unappealingly pithy for people but picked at by birds and iguanas, usually after it falls on the ground or in the water. Custard apple (*Annona reticulata*) (cachimán, cashema, chirimoya) and sugar apple (*Annona squamosa*) (anón, cachiman, pomme cannelle, sweetsop) are native species cultivated for their fruit, initially by Indigenous People. Soursop (*Annona muricata*) (graviola, guanabana, guanábano, korosò) is also an edible native species introduced widely elsewhere. These annonas are commercial fruit trees that also escape cultivation, establishing in scrub and previously cleared areas or persist at old home and garden sites. Annonas have many folk-medicine uses. Soursop is commercially available as an herbal medicine in powdered or pill form. Although sold as a cancer cure, there is no scientific support. Cultivated annonas are available in plant nurseries.

Lancewood (*Damburneya/Nectandra/Ocotea coriacea*) (avispollo, bastard torch, black torch, canela, cigua, laurel, pepper cillament, sweet torchwood, sweetwood, torchwood): This Caribbean native, related to laurels and bays, is a forest-understory tree growing to 10 m. It is prunable to a multi-stemmed shrub. Tolerating shade well, lancewood is an important addition to natural gardens. It hosts diverse insects including swallowtail butterfly caterpillars. Attractive, fragrant, clustered, spring flowers are sufficient for it to be considered a honey tree. Oil-rich, avocado-like, 1-cm fruit are available mostly in late summer-fall but also sporadically through the year, attracting fruit-eating birds especially mockingbirds, thrushes, and some warblers. Oily, hard, elastic wood was used for torches, shafts, and bows, and also for cabinetry. It is an essential element in a natural garden. It also is interesting in park settings and useful as buffer and in commercial settings. It is propagated by seed but better by transplanting wild seedlings. It is sometimes available in garden shops.

Avocado (*Persea americana*) (aguacate, alligator pear, avocat, midshipman's butter, pear, zaboka): Avocado is a nearly indispensable Caribbean yard tree for its nutritious fruit, so long as the yard is big enough. Originating in Mexico and Central America, it has been in cultivation for millennia. Reaching 15 m, it also is prunable to appropriate garden scales, which also is better for harvesting. As a commercial plant, it hosts many insects – ants, boring beetles, swallowtail caterpillars, lacebugs, leaf hoppers, mites, termites, thrips, wasps,

Lancewood.

and whiteflies. Insectivorous birds seem always to be hunting through its tight canopy of leaves, inflorescences, and branches. Flowers attract bats, bees, flies, and wasps. Woodpeckers and parrots open nearly ripe fruit that is then shared by other birds and insects. In good crop years, a gardener is likely to be willing to share some with birds, even if still on the tree. It is best propagated by grafting for faster production and predictable fruit.

Greenbriars (*Smilax*): These thorny, perennial vines can form thickets of painfully spiny branches running from ground to canopy. They provide shelter, nectar, and fruit for wildlife. Their small flowers are nectar sources; dark waxy fruit remain on the plant for a long time; and it will vine into the canopy providing structure and protection there. Native species occur throughout the Caribbean, 10 on the Greater Antilles alone. Greenbrier (*Smilax havanensis*) (chaney briar) is from The Bahamas, Caymans, Cuba, and Hispaniola. Wild bamboo (*Smilax auriculata*) (chinabrier) occurs in the Lucayan Archipelago. Bamboo vine (*Smilax laurifolia*) is from The Bahamas and Cuba. Basket vine (*Smilax guianensis*) (boyeu-chat, boyeau diable, liani-boyau) is from Puerto Rico and Lesser Antilles. Greenbriars' flowers draw bees, butterflies, and wasps. Dark, waxy, fruit persist until eaten by birds, which spread the seeds widely. Smilax tubers were a traditional source of food starch. Various species were used for medicine; and some properties have been proven. Introduced sarsaparilla (*Smilax ornata*) is in an herbal drink, called Baba Roots in Jamaica, which is used to promote health overall and stimulate sexual performance. Powdered roots are known as rad in Jamaica. In Cuba, pru Cubano is a fermented soda that includes zarzaparrilla (*Smilax domingensis*). Greenbriers are propagated by seed, cuttings, and tubers, but local species readily volunteer. Once established, the thorny vines resist assault and should be encouraged where wanted but perhaps discouraged where they are not.

Orchids (Orchidaceae): Orchids are interesting, aesthetically pleasing, and diverse in the Caribbean. They contribute 10% of the Caribbean's native plant species. Over 670 species are indigenous. *Encyclia, Epidendrum, Lepanthopsis, Maxillaria, Oncidium, Pleurothallis,*

Rich-flowering oncidium, a dancing lady orchid from a genus widespread in the Caribbean.

and *Vanilla* are among the better-known, native, widely dispersed orchid genera. Over 400 are endemic, and more are range-limited within the Caribbean. Epiphytic and vining species add structural and biotic diversity by providing living space for invertebrates. Flowers attract insects, usually bees, some of which depend on a single orchid species. There are many endemic and range-restricted species. Conservation of the Caribbean's native orchids and their pollinators is a high priority. One threat is poaching. Securing native orchids for restorations, taking care that it is legal, does not harm wild populations and can serve conservation purposes. Encouraging the propagation of native Caribbean orchid species for installations in restorations and natural gardens is a valuable conservation approach. A main wildlife value of orchids is adding structural diversity, this is one case where cultivated, nonnative species and hybrids can serve the same function. However, they do not support native pollinators.

Aloe (*Aloe vera*): Originally likely from the Arabian Peninsula, aloe's clustered, tubular flowers evolved for sunbirds but are used by bananaquits, hummingbirds, orioles, other nectar-seeking birds, and bats. Its clustered growth provides deep shelter for small animals. Aloe is well known for the sap of its fleshy leaves, which is famously used for skin care, hair, burns, digestion, and other health purposes. It does well in containers, which also constrain its spread. They may be propagated by seed but more easily by division.

Agaves (*Agave*) (aloe, century plant, daggers log, maguey, maypole, wild sisal): Agave of various species and appropriate size should be considered for every garden, especially in xeriscaping, but if for no other reason, for their sharp leaves that provide security for birds and other small wildlife and as a deterrent along boundaries. Species do differ in size, pointiness, and somewhat in structure. Native species occur throughout the Caribbean, some with restricted ranges. The Bahama agave (*Agave bahamana*), for example, is endemic to the northern Bahamas. West Indian dagger (*Agave kerrato/barbadensis*) (Antigua agave, century plant, maguey), from the Windward Islands, is the national flower of Antigua and Barbuda. Nonnative species include sisal (*Agave sisalana*), a Mexican native, once cultivated extensively for fiber through the region. The originally nonnative dagger's log (*Agave americana*) is cultivated worldwide. Agave bloom on tall spikes, some reaching tree-size, and provide nectar for bananaquits, bats (their major pollinators), butterflies, hummingbirds, and warblers. The plant dies after blooming. They are propagated from plants sprouting on the flowering stalk and division of pups. They can suffer from overwatering.

Bahama agave.

Spanish bayonet (*Yucca aloifolia*) (Adam's needle, aloe yucca, bayoneta Española, cucarachita, dagger plant, espadilla, Indian bread plant, Spanish dagger): Likely introduced to the Caribbean in colonial times and perhaps precolonial times from Central America, Spanish bayonet has naturalized through the region. It is naturally a salt-tolerating coastal plant that can reach 6 m but is usually about 2–3 m. Its dagger-like leaves provide impenetrable security and protection for birds, wildlife, and for people as a living fence. Attractive displays of white flowers are visited by nectar-seekers and pollinating moths that lay eggs in its flowers. Blossoms and fruit are edible, having been used by Indigenous People. It grows well in cultivation, and yuccas of several species are widely used in gardening. It is propagated by division from pups, and it is available from plant nurseries.

Palms (Arecaceae): Symbolically, culturally, economically, practically, and touristically, palms are important plants throughout the Caribbean. They provide food, homes, roofs, utensils, clothing, tourist interest, and wildlife habitat. Many species are used for thatching and weaving, often referenced in common names. The Caribbean is a center of diversification; over 75 species and 26 genera are indigenous to the region. Three genera are endemic, *Zombia*, *Calyptronoma*, and *Hemithrinax*. Most palms have small seeds eaten and spread by birds, and many have complex structure that shelters invertebrates and various small animals. Palms are used for nest sites by blackbirds, grackles, orioles, palm swifts,

Palm forest in Royal Palm Reserve in Jamaica, dominated by endemic Morass royal palms.

hawks, and owls and as roost and nursery sites by bats. Single-trunked plants provide nesting cavities; coconut and royal palms may be considered keystone plants for cavity nesters. Many palms retain their leaves or leaf bases, called boots, for some time, extracting nutrients and providing sites for a community of animals and epiphytes. Palms are ecosystems unto themselves. To provide this ecosystem, palm branches should not be trimmed, but left to fall in their species-specific way. Most palms bloom prolifically in summer attracting hordes of beetles, bees, and flies. Fruit-eating birds of appropriate size for each palm species consume the fruit. Palms do well in hurricanes, quickly resprouting new leaves. A typical tropical garden will include a volume of palms, albeit usually store-bought nonnatives. With such a wealth of indigenous palms, some of which are grown in gardens around the world, there seems hardly a reason not to only use native palms in parks, commercial plantings, and natural gardens.

Sabal palms (*Sabal*): Sabal palm (*Sabal palmetto*) (bay top, blue palmetto, cabbage palm, cabbage thatch palm, hat palm, pond thatch palm, pond top palm) occurs in Cuba and the Lucayan Archipelago. There also are 8 other sabal palm species in the region, which should be used locally when possible. Tall specimens provide perching and nesting sites for birds, including large owls and hawks. Single trunks provide nesting cavities. Bats roost under the leaves. It is host to skipper and owlet moth caterpillars. Unless removed, remnant leaf bases called boots remain on the trunks and capture debris creating microcosms of ferns, other epiphytes, ants, beetles, frogs, millipedes, spiders, tree lizards, worms, and weevils. Spring-summer flowers on huge stems attract bees, butterflies, flies, wasps, and nectar-feeding birds. Date-like, 1-cm, black fruit is eaten by birds, iguanas, and crabs and historically by Indigenous People – an acquired taste. They are propagated by seed and transplanting. *Sabal palmetto* is available in plant nurseries. Using island-indigenous sabal species provides conservation value.

Fan palms: Many species of native fan palms occur throughout the Caribbean. Most are very appropriate for natural plantings. They provide substantial structure, including by holding on to their old leaves or their bases for some time. All produce masses of flowers that attract nectar-seeking insects and fruit that attract fruit-eating birds and bats. The Caribbean has 50 species of *Cocothrinax* alone, including island-specific forms. The widespread silvertop palm (*Cocothrinax argentata*) (silver palm, silver thatch palm) is a short, 2–6-m, slow-growing, single-trunked palm. It is a coastal species primarily from Lucayan Archipelago and Greater Antilles. Caribbean thatch palm (*Thrinax radiata*) (chit, Florida thatch palm, Jamaican thatch, sea thatch palm, silk-top thatch palm) occurs in The Bahamas through Greater Antilles and a few in Lesser Antilles. Key thatch palm (*Leucothrinax morrisii*) (Bahamas broom palm, brittle thatch palm, brittle top palm, buffalo-top, guano de sierra, miraguano, palma de cogollo, palma de escoba, palmita, pandereta, yaray) is a tall, thin, single-trunked palm native to the Greater Antilles, except Jamaica, and the northern Lesser Antilles. Its many common names attest to its uses. The Cuban petticoat palm (*Copernicia macroglossa*) (jata de guanabacoa, jata palm) is a long-time horticultural favorite for interestingly holding onto its dead leaves, providing exceptional cover for insects, small birds, and tree lizards. These and more fan palms are excellent plants to consider for restorations, parks, and gardens within their native ranges.

Paurotis palm (*Acoelorrhaphe wrightii*) (cubas, Everglades palm, Madeira palm, papta, tique): Occurring naturally in The Bahamas and Cuba, this spiny, cluster-growing fan palm can reach 7 m, but mostly 3–4 m. Its sharp leaf stems provide security for birds and reptiles, deterring cats and people. Its small fruit is eaten by birds in its mature black stage. Its native habitat is slight elevations within flooded wetlands and so prefers damp locations, but away from concrete that leads to frizzletop syndrome. It is propagated by seed and division and is available in plant nurseries.

Hog cabbage palm (*Pseudophoenix sargentii*) (buccaneer palm, Sargent's cherry palm, Sargent's palm): This attractive, solitary, salt-tolerant, single-trunked palm, usually shorter

Hog cabbage palm with bananaquits foraging among the flowers.

than 5 m, is like a garden-sized royal palm. It is a coastal species occurring in The Bahamas, Hispaniola, Mona, Navassa, and Dominica. Three related species occur in Hispaniola. It grows in both sun and partial shade, preferring limestone substrates. Hog cabbage palm produces large inflorescences, most prolifically in summer attracting hordes of nectar-seeking insects. The larger fruit-eating birds consume fall-winter, red, 1–2-cm fruit. It can be readily grown from seed, which—unusual for a tropical plant—can withstand storage for a couple of years. It may be available in plant nurseries.

Royal palms (*Roystonea*): Royal palms are among the more important Caribbean wildlife plants. They are truly tall, towering to 30–40 m, sticking up above the normal forest canopy, and essential for restorations within their range. Palmchats and monk parakeets eat the flowers and build communal nests in them. Trunks, dead or alive, provide for cavity-nesting birds, especially woodpeckers and parrots, and even after the top has fallen off, the resulting cavity can be a nest site for barn owls. Despite their height, they also are garden and natural-area palms; at eye level in yard, a royal palm is only an attractive, steel-gray, lichen-covered column; the rest is overhead. The Caribbean has 10 royal palm species; Cuba alone has 5. The Morass royal palm (*Roystonea princeps*) is a Jamaican endemic. Cuban royal palm (*Roystonea regia*) occurs naturally in Cuba and Caymans but has been introduced widely elsewhere. Its spectacular height provides high perches; its large crown of live and dead leaves, spathes, and present and past inflorescences support communities of invertebrates, frogs, and lizards; and the flowers are typically covered with bees, butterflies, and flies. The fruit is special. One plant can produce 50 kg of fruit over a 6-month season, more than the local birds can eat, much of it piling up on the ground surrounding the trunk. On the tree, fruit attracts all the larger fruit-seeking birds; below it, every sort of fruit scavenger. Uneaten fruit piles add to the palm's soil base, elevating it above its usual moist ground. There are cautions. Falling leaves can be damaging and dangerous. Royal palms are not welcome over a car park or frequented sidewalk. Palm frond guards installed over precious plants can help. Roots are aggressive both below and above ground. Royal palms do well in nearly any planting situation from specimens to a diverse native garden. Every part of the plant traditionally has been useful. They are readily propagated by seed and by transplanting volunteer seedlings and even large plants. For conservation, species and plants indigenous to an island should be used; but plants have been shipped around for a century, so it is not always certain what the native stock is.

Coconut palm (*Cocos nucifera*) (coco, coconut tree, cocotero, el coco, noix de coco, palm de coco): How can a plant that is so much a part of Caribbean heritage not be native? Coconut palms came to the Caribbean from the Pacific by way of West Africa but have been a part of the Caribbean way of life for centuries. They have long been farmed commercially in orchards or groves for their coconuts but are equally at home decoratively in parks and yards. Trunks, alive or dead, are ideal for cavity-nesting woodpeckers, flycatchers, parrots, and parakeets. The broad crown of long leaves provides high perches and nesting sites for large birds. Bats and Bahama and Cuban orioles roost under leaves. The crown, when it is allowed by not cutting fronds early, holds its dead leaves for a time providing a community of earthworms, frogs, insects, lizards, millipedes, spiders, and other small animals that accumulate within the bases of the leaves, all of which are also food for other animals. Flowers attract bees and other insects. Coconut palms can be invasive on shores; but this is not a concern within the constraints of an inland site.

For birds, a succession of different-aged palms in a garden or park provides a continual supply of nest cavities. When one dies, if at all possible, its trunk should be kept standing for cavity-nesting birds. Propagation is from coconuts; just let them lay where they fall and then transplant if desired.

Sierra palm (*Prestoea montana/acuminata* var. *montana*) (chou palmiste, macoutouca, manacla, mountain cabbage palm, mountain palm, palma boba, palma de sierra, palmiste-montagne, palmiste rouge): Native to the Greater and Lesser Antilles, this tree-sized palm can reach 10–20 m tall. It has a thin, single trunk with buttress roots (the Antillean form of multi-trunked Sierran palm). It natively is a pioneer species occurring along mountain creeks, forest openings, and damp situations at lower elevation. It is a dominant plant in some Greater Antillean forests. Prop roots, epiphytes, and massive inflorescences provide exceptional structure. Flowers attract flies and bees, including honeybees. It produces thousands of 1-cm black fruit, which are eaten by strong-billed birds, especially by parrots but mostly fall to the ground for other animals or to become mulch. It is fast-growing and shade-tolerant and so is used in rainforest restoration and in parks and gardens. It is propagated by seed but is also available from plant nurseries.

Banana (*Musa* x *paradisiaca*): Originally Old World plants, bananas, and plantains are Caribbean food, home-garden, and commercial staples, grown throughout the region in yards, family farms, co-ops, and plantations. Historically, bananas were a dominating post-sugar crop, commercially and privately. They remain crucial both for domestic consumption and for export in some countries, such as the Dominican Republic. Hurricanes, disease, pesticide use, labor issues, and corporate and world economics and politics have been and are commercial issues. In a garden setting without pesticide application, banana plants, flowers, and fruit are wildlife-friendly. They provide considerable herbaceous structure and an ample supply of ants and other invertebrates. Leaves and fruit stalks attract ants, aphids, skippers and moths, spiders, thrips, weevils, and whiteflies. Flowers attract ants, bats, bees, bananaquits, hummingbirds, orioles, and other nectar-feeders. Ripened fruit attract bats, fruit-eating birds, butterflies, and moths such as the sphinx moth. A traditional way of growing bananas is clustering them together in available depressions, called banana holes. Emulating banana holes remains a good way to organize bananas in the confines of a garden as cluster-planting provides support against leaf-shearing and uprooting wind. On larger sites and in agriculture, plant clusters are arranged in rows according to local farming customs. Bananas need fertile, organic soil; in a garden, the banana hole is a good place for compost and manure. Although tissue culture is used commercially, bananas and plantains locally are propagated by transplanting pups.

Heliconia (*Heliconia caribaea*) (Caribbean heliconia, false bird-of-paradise, lobster claw, wild plantain): Native to Greater and Lesser Antilles, these large banana-like plants can grow to be 3–5 m tall. The flowers are tucked within showy red or yellow bracts that attract large hummingbirds. Tanagers and other fruit-eating birds consume the small blue-purple fruit. Heliconias are cultivated as ornamentals, so varieties are available in plant nurseries. They are a native substitute for the similar nonnative bird-of-paradise.

Bromeliads (Bromeliaceae): Bromeliads are pineapple-like plants. Numerous epiphytic species are native to the Caribbean, mostly the genera *Tillandsia, Catopsis*, and *Guzmania*. They tend to occur in relatively moist forests with breezes and tend to be where they can get light. There are many more horticultural nonnative species and varieties. In some jurisdictions, native species can be rescued from development. As air plants, bromeliads add substantial structure providing habitats and microcosms for communities of insects, frogs, and lizards. Vase-shaped bromeliads create aquatic microcosms in their "tanks" providing dry-season water for birds, frogs, and lizards. Bromeliad flowers attract and are one of the most important nectar sources for hummingbirds. They also attract bats and bees. Management of bromeliad tank mosquitoes is discussed in Chapter 4. One of the more widespread vase bromeliads is wild pine (*Tillandsia utriculata*) (giant airplant, giant wild pine, spreading airplant, swollen wild pine). It is hardy as an epiphyte in various sorts of habitats. Spanish moss (*Tillandsia usneoides*) also is widespread through the region, providing unique habitat within its branch-draping stem masses. Pinguin (*Bromelia pinguin*) is a widely planted Caribbean native. Commercial pineapple (*Ananas comosus*) is a terrestrial species that is considered native as it was introduced from South America by Indigenous People but became a favorite in colonial Europe and a symbol of Jamaica and Antigua. Its fruit and its uses are well known. As a garden plant, it provides moderately thorny and skin-irritating low-stature cover. There are many cultivated varieties in the Caribbean such as Antigua black, ripley, smooth cayenne, Spanish red, and sugar loaf. Although single-blooming, pups are transplantable, and fruit tops can be planted. Native bromeliads are dispersed through wind-blown seeds and so will volunteer. Adult plants can be tied to appropriate trees.

Grasses (Poaceae): Grasses are one of the largest plant families with species occurring naturally in nearly all Caribbean habitats, and more introduced. Recovering Caribbean nature requires encouraging native grasses while discouraging invasive nonnative species, which are no trivial tasks. Some nonnative grasses were imported to improve pastures;

Tillandsia bromeliad being explored by a Jamaican woodpecker.

others were accidental stowaways. The most common use of imported grass is for pasture and lawns. The values of replacing sod monoculture lawns with diverse lawns are discussed in Chapter 4. Tall-stature, ornamental grasses provide cover. Small seed-eating birds, such as buntings, grassquits, and sparrows, directly consume the seeds that grasses produce. All grasses support their diverse complement of insects including beetles, small flies, grasshoppers, small moths, spiders, and worms, which can become bird and lizard food. Flycatching birds tend to forage on insects over grass; and other species, such as mockingbirds, blackbirds, grackles, and some sparrows and thrushes, feed preferentially by walking or hopping through grassy areas.

There are many grass species in the region, both native and nonnative. Commonly used, nonnative species include muhly grass (*Muhlenbergia* especially *Muhlenbergia capillaris*); pampas grass (*Cortaderia* especially *Cortaderia selloana*); reed canary-grass (*Phalaris arundinacea*) used in landscaping; African Bermuda grass (*Cynodon dactylon*) commonly used as lawn grass; West Indian lemon grass (*Cymbopogon citratus*) used as a herb, despite its name actually from Southeast Asia; Guinea grass (*Megathyrsus maximus*) (green panic grass) a drought-resistant, shade-tolerant, fire-prone, African species that invades sugarcane; and African signal grass (*Brachiaria decumbens*) used for pasturage. Native grasses can instead be used in gardens and restorations, including broom grasses (*Andropogon*) (beard grass, bluestem grass, broomsedge), gama grasses (*Bouteloua*), and Caribgrass (*Eriochloa polystachya*) (Caribbean cupgrass, malojilla, paragrass, pasto caribe, pasto janeiro, yerba de Caribe). There are many more that could be taken into horticulture. Grasses are propagated by seed and division.

Grapevine (Vitaceae): Grapevines provide structure and shelter within the canopy and, of course, grapes. Their flowers attract bees and other insects. Fruit-eating birds, iguanas, and rats take grape fruit avidly. Native West Indian grape (*Vitis tiliifolia*) (Caribbean grape, water vine) is a vigorous climber reaching 35 m. Sorrelvine (*Cissus trifoliata*) (hierba del buey, possum-grape, vine-sorrel) is a hardy pioneer on cleared land. Historically, grapes were indigenous forest crops. They are propagated by seed, cuttings, and air layering and will volunteer.

Bay cedar (*Suriana maritima*) (cedar, cuabilla, cucharo, guitarán, jennifer tree, jovero, junifer, juniper, krist marinn, oseille bord de mer, palo corra, pantsil, perce-pierre, romarin noir, temporana, tassel plant): This pantropical, native, coastal shrub is an excellent candidate for a natural or xeric garden as well as for commercial applications. It is a 2–3-m shrub, rarely taller, with a short, very heavy trunk, gracefully arching branches, and dense, thick, evergreen leaves on short, upturned twigs producing a densely rounded crown. They are flexibly used in containers, as a shrub, as a single-species or diverse hedge, as buffers, and in commercial settings such as parking lots and median strips. Bay cedar provides excellent cover and supports numerous insects including hairstreak caterpillars. Inconspicuous, yellow, branch-tip flowers attract bees, butterflies, and other insects. Small 1/2-cm fruit are consumed by fruit- and seed-eating birds and iguanas. The slightly cedar-scented leaves are used as a tea for intestinal disorders and fevers, as a strengthening tea (and mariner's poor coffee substitute), and also for cleansing sores; and bark, for toothaches. Because they are coastal plants, they are hardy and drought-tolerant. Bay cedar is propagated by seed and hardwood cuttings.

Caesalpinias (Caesalpinioideae): Species of caesalpinias are fast-growing, weedy shrubs to small trees with dense spiny branches that provide shelter and bird nesting sites. They are hosts for blue butterflies. Their spectacular flowers attract hummingbirds and butterflies, and their seeds are widely eaten by birds. Several well-known species are native. Brasiletto (*Caesalpinia vesicaria*) (die wood, large-leaved cassia) was harvested as a red dye source and now is protected in some jurisdictions. Divi-divi (*Caesalpinia/Libidibia coriaria*) (cascalote, guaracabuya, guatapana, nacascol, watapana) occurring in The Bahamas, Greater Antilles, and Leeward Antilles is famous for being shaped by coastal wind; it is the national tree of Curaçao. Several species of nickerbeans (*Caesalpinia/Guilandina*) are viny, coastal, seabean-producing shrubs that provide superior thorny shelter. They are propagated by processed seed. Some caesalpinia species are available in plant nurseries.

Sennas (*Cassia/Senna*): Sennas are multi-stemmed, sprawling, leguminous shrubs that are hardy, weedy, fast-growing, and salt-tolerant, but short-lived. They provide structure, insects, flowers, high-quality seeds, and decoration in gardens, parks, and natural plantings. Senna canopies host ants, spiders, sulphur butterfly caterpillars, and many other insects. Striking flowers attract ants, bees, and butterflies. Seedeaters, especially bullfinches, take seeds from pods. Well-known sennas are Chapman's wild sensitive plant (*Senna mexicana* var. *chapmanii*); mosquitobush (*Cassia biflora*) (desert cassia); candlebush (*Senna alata*) (candelabra bush, candletree, casse allee, casse puante, Christmas candle, dartier, impetigo bush, tarantana); and butterfly bush (*Cassia/Senna bicapsularis/emarginata*) (black dog bush, blydog, bois d'anneau, caca-betye, Christmas bush, money bush). Nonnative yellow cassia (*Senna siamea*) is widely planted streetside, and the nonnative golden shower tree (*Cassia fistula*) is used as a specimen tree. Sennas are propagated by seed, cuttings, and transplanting very young seedlings.

Bay cedar with a wintering black-throated green warbler.

Chapman's wild sensitive plant.

Guaba (*Inga vera/edulis/feuillei*) (cuaniquil, guaba nativa, guama, guama jina, guavo, ice cream plant, inga, ingá verdadeiro, joaquiniquil, pacay, pan chock, pois doux, pois doux à paille, pois sucrin, pwa dou, river-koko, sikren): This widespread, Caribbean native legume has been cultivated for centuries. It is a highly variable, genetically diverse plant problematic for taxonomists. As a fast-growing tree of 4–12 m tall, it readily colonizes waste places. Its year-round flowers attract its main pollinators, bees, and also other nectar-seeking insects and birds. Seed pods, called ice cream beans, have smooth sweet pulp that is edible by people and birds. It is planted for shade over coffee and cacao. Guaba is propagated by seed, and cultivars are available in plant nurseries.

Lysilomas (*Lysiloma*): Common native pioneer leguminous species, lysilomas, make excellent native wildlife plants. They are spreading shade trees, up to 10 m tall, that provide mottled light through leaves in the wet season and are dry-season deciduous. Flowers, flaky bark, leaves, and seed pods host tree snails and insects including ants, termites, thornbugs, and caterpillars including sulfurs, yellows, hairstreaks, and blues. Flycatchers, gnatcatchers, vireos, and migrating warblers sort through branches for insects. Showy, 1-cm, fragrant, powderpuff flowers attract an array of butterflies and moths. They fruit in summer-fall. Grackles, crows, and other birds eat the seeds once the flat pods open. Wild tamarind (*Lysiloma latisiliquum/bahamense*) (Caribbean walnut, false tamarind, horseflesh, lysiloma, sabiccu, sabicú, taverneau, t'zalam) occurs in The Bahamas and Cuba. Horseflesh (*Lysiloma sabicu*) (bacona morada, sabucu, tabèno, teverneau, wild tamarind), occurring in The Bahamas, Cuba, and Hispaniola, was heavily harvested for use in boats and furniture. Lysilomas are tolerant of various soil types and storm resistant. They are propagated by seed and transplanting seedlings.

Wild tamarind.

Pithecellobiums (*Pithecellobium*): Species occur throughout the Caribbean and are all wildlife-friendly. Large-growing species have long been used as specimen plants in public gardens. Leaves host butterfly caterpillars, including sulphurs, blues, and skippers. They have a protracted blooming season during which showy, powderpuff flowers attract honeybees and butterflies. Long seed pods open spectacularly to expose pulp (aril) that are hugely attractive to birds. People also eat it. Blackbead (*Pithecellobium keyense*) (keys blackbead, ram's horn) is a shrub to small tree from the Lucayan Archipelago and Cuba. Blueseed (*Pithecellobium/Abarema glauca*) (caracolí, jijaguao, merey montañero, nijaguao, paují) from The Bahamas and Greater Antilles, The Central American pinzán (*Pithecellobium dulce*) (camachile, Madras thorn, monkeypod tree), is widely introduced for its interest and pods with sweet edible pulp. It is cultivated in horticulture for its attractive flowers and similarly attractive coiled fruit. Cat's-claw (*Pithecellobium unguis-cati*) is a thorny shrub to small tree that provides exceptional shelter and is especially suitable for diverse hedgerows. Other native species occur but are more range-restricted. Pithecellobiums were traditionally used for their lightweight wood, tannin, fruit, and seeds as food, and for medicinal purposes. Nitrogen-fixers, they are valuable early in reforestation. Propagation is by seed, cuttings, and air layering.

Acacias (*Vachellia/Acacia*): Acacias are thorny legumes with interiors that provide exceptional security and nesting sites. They are quite hardy and adaptable. Acacias have attractive, fragrant, spring-summer, powderpuff flowers on stalks that attract bees, butterflies, flies, and wasps. Thick, 4–6-cm pods break open exposing white pulp and seeds that birds consume. Native acacias are suitable for use in nature recovery from ecosystem restorations to gardens but can be invasive outside their native ranges. Sweet acacia (*Vachellia farnesiana*) (huisache, needle bush, and other names in other places) is a Caribbean-wide native, although with a wide natural and introduced range. Rosewood (*Vachellia acuifera*) (cassip) occurs in The Bahamas and Cuba. Cinnecord (*Vachellia choriophylla*) (tamarindillo, thornless acacia) is a thornless species occurring in The Bahamas and recently in

Cuba. It is usually a multiple-stemmed shrub or small tree, up to 10 m tall. The curled, spreading, brittle branches are prunable into a hedge or as part of a diverse hedgerow. Acacias are propagated by seed and cuttings and are available from plant nurseries.

Dogwood (*Piscidia piscipula/erythrina*) (fishfuddle, fishpoison tree, Jamaica dogwood): This Caribbean-wide native is usually a medium-sized tree, although potentially reaching 10–17 m. It is fast-growing, blooming within a few years. Despite rotenone-infused defenses, its foliage supports insects including skipper, blue, and hairstreak caterpillars. A far-reaching canopy of big leaves provides cover. In spring when the plant is leafless, its attractive white to pink, clustered, pea-like flowers cover the tree attracting ants, bees including honeybees, bananaquits, butterflies, hummingbirds, and tanagers. It is quite striking at this stage. It was historically used for stunning fish. Its hard wood was traditionally used for boat framing, poles, posts, and charcoal. It is propagated by seed, which germinate in only a few days, or by cuttings.

Butterfly pea (*Centrosema virginianum*) (blue bell, spurred butterfly pea, wild blue vine, wild pea): This delicate, twining, thin-stemmed, leguminous vine is common throughout the Caribbean and beyond, particularly in sandy soils. It is fast-growing and drought-tolerant. Its foliage hosts cloudywing and skipper butterfly caterpillars. It can trail along ground or climb through shrubs, attractively filling space. Prolific, purple to light blue, summer flowers attract bees. It is easily propagated by seed.

Gliricidia (*Gliricidia sepium*) (Aaron's rod, fast stick, madero negro, madre de cacao, mata ratón, Mexican lilac, quickstick, piñón de Cuba, piñón florido): Likely a Mexican/Central American native, this useful, dry-season deciduous, 10-m tree with a spreading crown and sensational dry-season flowering was indigenously cultivated and introduced pantropically, including through the Caribbean. Its prolific, pink-white, flowers at the

Jamaica dogwood tree and female scarlet tanager.

end of branches are readily available for nectivores, including bees, especially large species such as carpenter bees, blackbirds, hummingbirds, orioles, bullfinches, and some warblers, which flock to it in turn. It is fast-growing, has a wide soil tolerance, and will readily self-seed sufficient to be considered a pest in some situations. It has deep roots, occupies slopes, and tolerates repeated cutting. Its many uses include fire wood, fodder, shade, trellis, living fences, erosion control, nitrogen fixation, including as pioneers in reforestation. It is propagated by soaked or scarified seed or cut branches.

Tremas (*Trema*): Hardy, pioneer plants of disturbed sites, roadways, scrub, and pine forests, some native tremas have large Caribbean ranges. This adaptability makes them good garden plants, particularly as edge plants along natural areas. These are shrubs or short trees, up to 6 m tall; they usually have an open growth form of thin erect branches. Abundant tiny flowers clustering along slim branches attract bananaquits and other nectar-seekers. The 1 mm pink fruit stays on the plant for long periods and is eaten by bullfinches, grassquits, mockingbirds, thrashers, some warblers, and vireos. Native species include West Indian nettletree (*Trema lamarckiana*) (bo hog bark, bois mahaut, cabrilla, Lamarck's trema, mahaut piment, pain-in-back, West Indian trema) and Jamaican nettletree (*Trema micrantha*) (bois chique, bois de scie, guacimilla). Leaves and bark are used in various cultures for treating diverse medical issues including diabetes, respiratory and urinary problems, toothaches and other pain, worms, and malaria, some of which are supported scientifically. As a pioneer species, they need open sunlight. Tremas grow fast and are easily started from seeds.

Lignum vitae (*Guaiacum sanctum* and *Guaiacum officinale*) (bois de vie, bois saint, brazilwood, gaïac, gaiac blanc, gaiac cardasse, gayak mal, guaiacum, guayacan, guayaco,

Lignum vitae (*Guiacum sanctum*).

holywood, ironwood, palo santo, tree of life): Sanctum occurs in The Bahamas and Greater Antilles tending to be in forest and rocky situations. Officinale is Caribbean-wide and typically more coastal. Slow-growing, those found today on Caribbean islands are usually short trees growing to about 4 m. They are widely appreciated for their flowers, fruit, history of use, and attractive rounded crowns of small, dense leaves that provide cover and nest sites. They bloom in spring-early summer, their pretty blue flowers attracting bees, butterflies, other insects. The striking, yellow-orange fruit with red seeds are consumed by fruit-eating birds including flycatchers, grackles, mockingbirds, orioles, parrots, tanagers, and others. Officinale is Jamaica's national flower; sanctum is Bahama's national tree. The persistent, dense, resinous wood historically was used for its strength, denseness, and oiliness. As a result, lignum vitae was heavily harvested throughout Caribbean. Resin has been used for cuts and bruises, leaves as a tonic for numerous ailments including gout and arthritis, and flowers and bark as laxatives. They are hardy and tolerant of salt, poor soil, and dry seasons. They are propagated by seed, transplanted as seedlings, and available at plant nurseries.

Black ironwood (*Krugiodendron ferreum*) (bois de fer, bois fait, bois petites feuilles, leadwood, palo de hierro, strongback): Found Caribbean-wide, this narrow-crowned shrub or tree growing to 10 m supports an abundance of invertebrates that attract insect-eating birds and reptiles. Tiny, 1-mm, yellow-green flowers attract insects; and birds eat the summer, black, 1-cm fruit as quickly as it becomes edible. Leaf tea is used for backache. Slow-growing, black ironwood wood is among the densest known. It is propagated best by transplanting seedlings.

Darling plum (*Reynosia septentrionalis*) (red ironwood): Occurring in The Bahamas and Cuba, darling plum is usually a tree up to 7 m tall but can be kept as a shrub by pruning. It is hardy, growing in nearly any sort of soil condition, and so does well in restorations, parks, or gardens. It has a slender trunk, stout branches, and narrow crown that can fit in tight places adding to cover in a diverse natural garden. It blooms all year but concentrated in spring and early summer with inconspicuous yellow-green 1/2-cm flowers. Birds and people eat the sweet, greenish pulped, dull-black, 1-cm fruit. Darling plum is propagated by seed and cuttings.

Colubrinas (*Colubrina*): Having attractive patchily peeling bark, colubrinas' thick branching provides exceptional wildlife structure and cover. They can be shade trees, ornamentals, or part of a diverse natural garden. Colubrinas bloom nearly year-round having tiny yellow flowers that attract large numbers of insects preyed upon by insectivorous birds. Tiny, 5-mm, prolific, capsule fruit are tightly attached to stems and release seeds by exploding (hence one of the names, soldierwood). Prior to the explosion, the fruit attracts seed-eating birds. Native species include soldierwood (*Colubrina elliptica*) (amole, behuco indio, blackbead, bois mabit, bwa mabi, coffee colubrina, mabi, mauby, nakedwood, snakebark) and coffee colubrina (*Colubrina arborescens*) (bijaguara, bois capable, bois de fer, bois mabi, bois pele, bwa ple, corazon de paloma, greenheart, soaptree). Both are common, Caribbean-wide natives. Soldierwood trees can get to 15 m or more, whereas coffee colubrina is usually a shrub or tree up to 7 m tall. They are used for skin issues, and the bark and leaves are made, at home or commercially, into a drink, mauby (mabi, maubi, mavi), used to increase stamina. Colubrinas are propagated by scarified seed and transplanting.

Figs (*Ficus*): Native figs are keystone wildlife plants throughout the Caribbean. They provide exceptional structure for wildlife, support numerous spiders and insects including daggerwing butterfly caterpillars, and the 1–3-cm fruits have high sugar and protein content. They are consumed by any and all of the fruit-eating resident, wintering, and migrant birds; a short list includes regulars such as bats, catbirds, doves, some flycatchers, especially kingbirds, mockingbirds, pigeons, thrushes, vireos, some warblers, and some woodpeckers. They also are eaten once on the ground by birds, insects, tortoises, and iguanas. Individual trees fruit at different intensities, at different times, and in different years, diversifying availability. Fig trees, including native figs, can get huge, 10–15 m and wide-spreading, but, if managed early, can be kept pruned to nearly any size and shape. Strangler fig (*Ficus aurea*) (fig, golden fig, higuerón, longleaf fig, West Indian fig) is native to The Bahamas and Greater Antilles. West Indian laurel fig (*Ficus americana*) (Jamaican cherry fig) occurs Caribbean-wide. It shows strong genetic diversification, so it is important to use stock indigenous to an island. Shortleaf fig (*Ficus citrifolia*) (aralie cerise, banyan, bearded fig, figuier, giant bearded fig, wild banyan, wild banyan tree, wild fig) also occurs Caribbean-wide. This species is best for small areas as it does not banyan. Nonnative figs are rather common in the Caribbean including the rubber plant (*Ficus elastica*) from Asia, weeping fig (*Ficus benjamina*) from Asia and Australia, Indian banyan (*Ficus benghalensis*), and sacred fig (*Ficus religiosa*) from the Indian subcontinent. These nonnatives can exhibit yard-covering, deeply shading, foundation-threatening banyan growth. They are not environmentally helpful. Native figs of indigenous island stock need to be used in nature recovery, and nonnative figs need to be replaced where possible in gardens and other settings. Native figs are propagated by seed, transplanting seedlings, cuttings, and air layering; and they volunteer.

Strangler fig with a Cape May warbler eating fig fruit.

Cecropia and Jamaican euphonia.

Trumpet trees (*Cecropia*): Prominent, distinctive, and naturally abundant trees, cecropia are also valuable wildlife plants. Reaching 20 m on a narrow stem, they sport umbrella-size and -shaped leaves. They are fast-growing, pioneer species on disturbed sites and so do well in parks and gardens. They are valuable in forest restorations after wind damage. The fleshy fruit – known as snake fingers and sometimes gummy-bear fruit, suggesting shape and taste – is eaten by fruit-seeking birds and bats, which spread the tiny seeds. *Cecropia schreberiana* (cecropia, pumpwood, guarumo, trumpet tree) is from the Greater and Lesser Antilles. *Cecropia peltata* (bacano, bwa kannon, cecropia, pumpwood, snakewood, trumpet, trumpet bush, trumpet tree, yagrumo hembra) occurs in Jamaica within the region but has an extensive natural and introduced distribution. It is highly invasive where introduced. The light wood and hollow branches have many practical uses. Cecropia are propagated by recent seed germinated in the sun.

Bayberry (*Myrica/Morella cerifera*) (arbol de cera, candleberry, cerero, forgotten tree, palo de cera, tallow shrub, waxberry, wax myrtle, waxwood): Bayberry is a critical bird-plant. Due to its nitrogen-fixing symbionts, it is a fast-growing, shrub or small tree up to 3–7 m tall. A native in the Caribbean, it has been introduced to gardens worldwide. It can be used in clusters, pruned to a small tree, made a visual-barrier hedge, or used in a diverse natural garden. Its spreading crown formed by a thicket of thin branches creates closed internal space that provides well-used cover. It is host for caterpillars of dusky wing and hairstreak butterflies. The tiny flowers are wind pollinated. In fall, female plants produce numerous, 1-mm, blue-green fruit with energy-rich waxy coats that persist into winter and are held even after hurricanes when other fruit becomes unavailable. The fruit attract fruit-eating birds including thrushes, kingbirds and vireos, and especially yellow-rumped, palm, and other wintering warblers, and wintering catbirds just before their spring migration. Leaves and fruit are used for flavoring and extensively in herbal medicine. As a pioneer species, it is hardy in natural gardens and commercial plantings but may do poorly in alkaline soil. Bayberry is best propagated by root suckers and cuttings and is available from plant nurseries.

Christmasberries (*Crossopetalum*): Two native species have quite different structure, but both have 1/2-cm, red, summer-fall fruit that attract birds. Maidenberry (*Crossopetalum rhacoma*) (cherry, poison cherry, rhacoma, sewal, tsewai, wild cherry), widely dispersed in the Caribbean, is a small tree growing to 5–8 m but usually shorter. Its thin drooping branches create fountain-like, open-branching inner structure providing cover. Its mostly spring-summer flowers attract bees, butterflies, and hummingbirds. As a forest and scrub plant, it does well in mixed-species settings. Quailberry (*Crossopetalum ilicifolium*) (Christmasberry), from The Bahamas, Cuba, and Hispaniola, is a short-statured, ground cover to 20 cm tall with small holly-like, spiny leaves. Both species are propagated by scarified seed and sometimes available from plant nurseries.

Maytens (*Maytenus*): Attractive, especially when in fruit, maytens are shrubs or small trees. Small, thick, highly prunable leaves provide cover for invertebrates, birds, and lizards. The tiny flowers attract bees. In summer and fall, the 1/2 to 1-cm, orange, capsule fruit attract fruit- and seed-eating birds. Several occur in the Caribbean. Boxwood (*Maytenus buxifolia*) (aguacero, boxleaf mayten, bwa foumi, carne de doncella, maytene a feulles de buis, rockwood, sangre de torro, spoon bush, spoonwood), occurring in The Bahamas, Cuba, Cayman Islands, and Hispaniola, is a shrub or tree reaching 10 m. Caribbean mayten (*Maytenus cymosa*) (bastard chelamella) is from Puerto Rico and the Virgin Islands. Maytens have long history in folk medicine, and hundreds of unique chemicals have been extracted from these plants; one now is synthesized for its antitumor and antibiotic properties. Extract from the Brazilian espinheira santa (*Maytenus ilicifolia*) is available commercially for stomach issues. Caribbean maytens are adapted to dry scrub and so are hardy and thrive in plantings. They are propagated by seed.

Clusias (*Clusia*): Hundreds of clusia species occur throughout the American tropics. Their thick-leaf canopy provides exceptional cover for animals. Clusia flowers have a variety of pollinating strategies; most, but not all, attract insect pollinators. The capsule fruit persists well into the winter when other food becomes scarce. When open, or opened by parrots and woodpeckers, the jelly-like pulp is consumed by fruit-eating birds and bees. In the Caribbean, indigenous and some endemic clusias may be found from seashore scrub to mid-altitude forests. Various species are shrubs, trees, vines, or host-strangling epiphytes. Clusia trees dominate some Greater Antillean montane forests. Two common species are balsam apple (*Clusia rosea*) (autograph tree, bois pale, bwa pal, copey, figuier maudit, pitch apple, Scotch attorney, wild mamee, wild mamee apple) and tar gum tree (*Clusia minor*) (autograph tree, copey clusia, cupey de monte). Both are widespread in the region and also are in cultivation. Balsam apple can become a tall, straight tree, growing up to 10 m, but more often is pruned to a thick hedge, although hedging reduces fruiting. Balsam apple is best as a shrub or tree. Tar gum can start out as an epiphyte. Nonnative *Clusia fluminensis* from Brazil is also used as a hedge but can be invasive and is best avoided. Common clusias are overused as monospecific hedges, most of which are male plants and so do not provide fruit for wildlife. But other indigenous clusias deserve conservation attention. Clusias produce an array of phytochemicals that are used variously

Balsam apple.

in folk medicine. Amazingly hardy, nearly indestructible once established, native species are valuable in mixed natural plantings. They are propagated by seed, cuttings, and air layering. Balsam apple is available from plant nurseries.

Coco trees (*Erythroxylum*): Over 30 erythroxylum species occur throughout the Caribbean. They are shrubs to small trees; some are soil-restricted. Branches and leaves provide cover and are host to moths. Small, whitish to greenish, 1-cm clustered flowers on small stalks attract bees, flies, moths, and wasps. Oblong, small, 3/4-cm red fruit are eaten readily by a diversity of birds including crows, pigeons, thrushes, and vireos. Examples include false cocaine (*Erythroxylum areolatum*) (smokewood, swamp redwood), ratwood (*Erythroxylum rotundifolium*) (swamp redwood), and jibá (*Erythroxylum havanense*). Propagation of native, noncultivated species is by seed and cuttings.

Locustberry (*Byrsonima lucida*) (candleberry, clam cherry, gooseberry, guanaberry, plumberry): Widespread in the Caribbean, locustberry is usually a thick, flat-topped, multi-stemmed, 2–3 m shrub but can be a medium-sized tree up to 10 m tall in appropriate soil conditions. It requires full sun and well-drained soil and is drought-tolerant. It is host for duskywing butterflies. In spring and early summer, showy, 1-cm flowers in impressive terminal clusters cover the plant. Flowers individually start white, turn pink, then red, attracting different butterflies and also oil-collecting bees. In summer and early fall, edible (but soapy), 1/2 to 2-cm, red-brown fruit attract all the fruit-eating animals and are persistent on the plant. There are other byrsonimas in the Caribbean worthy of local conservation concern. The bark has been used in folk medicine mostly for treating respiratory and dysentery-like issues. Propagated by seed and by transplanting volunteers. It is also sometimes available from plant nurseries.

Locustberry.

Wild cherries (*Malpighia*): Some wild cherry species are similar and often confused, so similar names are applied across species and ranges, including acerola, Barbados cherry, cereza, cerise de Saint Domingue, garden cherry, grosella cerisier, manzanita, native cherry, seriz, shimarucu, touch-me-not (for some stinging species), West Indian cherry, and wild crepe myrtle. They are spreading shrubs growing up to 4 m with multiple stems and small leaves that create excellent cover within their open interior. Flowers attract bees and butterflies. Bright-red 2-cm fruit, available from summer through winter, are consumed in pieces by larger fruit-eating birds. Some native species are *Malpighia glabra/punicifolia* from the Greater and Lesser Antilles; cultivated *Malpighia emarginata* (acerola, Barbados cherry, cerise de Saint Domingue, guarani cherry, ti seriz petit cerise, West Indian cherry); *Malpighia polytricha* from The Bahamas and Hispaniola; stingingbush (*Malpighia setosa*) from Puerto Rico; and palo bronco (*Malpighia fucata*) from Puerto Rico, Virgin Islands, and Cuba, which causes dermatitis; but a dry branch or bronco stick is protection against all sorts of misfortune. Emarginata fruit are well known and commercially farmed as a vitamin C source. Wild cherries are propagated by seed, cuttings, and air layering. Species, mostly emarginata, are available in plant nurseries.

Cocoplum (*Chrysobalanus icaco*) (abajeru, cocoes [fruit], icaco, icaque ponne, ikak, fat pork, paradise plum, pork fat apple, zicate): Widespread in the Caribbean and beyond, this flexible plant can be a shrub, windbreak, hedge, or a tree of about 6 m, essentially prunable to whatever might be wanted. Its dense foliage of dark-green leaves is attractive and functional, providing cover for numerous invertebrates, especially ants and spiders,

Cocoplum.

as well as warblers and vireos and lizards hunting them. Flowers attract bees and butterflies. Plants fruit at various times, mostly summer. Large, 2–3-cm, purple, pink, or white fruit are eaten by large birds and iguanas on the tree and by others on the ground. They are eaten by smaller birds in bits. The white cottony flesh and seed are quite edible by people; the fruit makes excellent jelly. Cultivars include "horizontal" a ground cover and "red tip," retaining juvenile leaf color. Taller shrubs with lots of fruit are most beneficial to birds. Cocoplum is propagated by seed and cuttings kept damp. Cultivars are available in plant nurseries.

Passionvines (*Passiflora*): One of most important vines for natural gardens, primarily as a butterfly host for heliconius, fritillary, and Julia butterfly caterpillars. These and other butterflies also are attracted to their flowers along with bananaquits, bats, hummingbirds, honeybees, bumblebees, carpenter bees, and wasps. Small fruit is eaten by all fruit-eating birds, spreading native species widely. Passionvines are climbers and can take over a tree canopy, fence, or an entire roof. Native species are particularly hardy. Corkystem passionvine (*Passiflora suberosa*) (juniperberry passionvine, small passion flower) and red-flowered passionvine (*Passiflora rubra*) are region-wide natives. The beautiful wild watermelon (*Passiflora cupraea*) (devil's pumpkin) is a native in The Bahamas, Cayman Islands, and Haiti. Other species are more localized. Also grown in the Caribbean are non-native, edible species: large-fruited barbadine (*Passiflora quadrangularis*) (badea, giant granadilla, giant tumbo, grenadine); edible passionfruit (*Passiflora edulis*); maypop (*Passiflora incarnata*); and sweet granadilla (*Passiflora ligularis*). Passionvines are propagated by seed, best planted immediately with pulp, and also by cuttings and air layering. Local species volunteer. Edible-fruit plants are available in plant nurseries and online.

Wild watermelon passionvine.

Yellow alder.

Alders (*Turnera*): Caribbean alders have prolific 2-cm yellow, bell-shaped flowers that attract hummingbirds as well as bees and an abundance of butterflies. They will grow almost anywhere in sand, rocky lands, pots, or on rocks and are quite dry-season tolerant. Yellow alder (*Turnera ulmifolia*) (buttercup, cup-and-saucer, Miss Williams, rangoat dashalong) and damiana (*Turnera diffusa*) (damiana, thé bourrique, Mexican holly, old woman's broom) are widespread through the Caribbean. Damiana has been used in folk medicine for centuries as an aphrodisiac, for menstrual pain, as a fish poison, and in Mexico as a liqueur. Leaf extract is available commercially, marketed as an aphrodisiac. Alders are propagated by seed, cuttings, and by transplanting volunteer seedlings.

Kase sek (*Samyda pubescens/Casearia hirsuta*) (wild coffee): This Greater Antilles native is an attractive, fast-growing shrub up to 4 m tall, although can get taller. It is salt- and cold-hardy, occurring on diverse soils but mostly on limestone. The soft leaves create cover. Clusters of fragrant, pink flowers make it an attractive garden plant and are visited by bees. The 2-cm capsule fruit incorporates a fleshy aril consumed by fruit-eating birds. This hardy species not only is valuable in restoration but also can be grown in containers. It is propagated by seed and may be available in plant nurseries.

Bull (*Pera bumeliifolia*) (black ebony, brown ebony, bullwood, ebony, hiki, jiqi, lipper plant): Occurring in The Bahamas, Cuba, and Hispaniola, bull is a beautiful 7–12 m tree having a slender, dark trunk and thick lanceolate leaves. Small, creamy, spring-summer flowers attract bees and other insects. In summer through winter, clusters of olive-shaped, 1-cm fruit on female trees produce a red pulp that birds eat as fast as it is available. Bull was previously harvested almost to extinction for its hard wood. As a pioneer tree, it is suitable for restoration and other natural plantings. It can be propagated by seed collected before fruit opens and planted immediately in the shade and by cuttings.

Crabwood (*Gymnanthes lucida*) (aceitillo, bois marbré, bwa mabre, crab bush, false lignum vitae, goat weed, granadillo, oysterwood, potatowood, shiny oysterwood): Widespread in the Caribbean, it is a tree, 3–8 m, with a narrow, open crown of distinctively upright branches. It flowers and fruits throughout the year, concentrated in spring. Fragrant yellow-green flowers attract insects. Small, 1-cm, 3-lobed, brown to black fruit hang on relatively long stalks where they are accessible to fruit-eating birds. It is a scrub plant and so is hardy and adaptable in planted situations. Crabwood is widely used in folk medicine for treating mouth and stomach issues as a tea or by chewing leaves or twigs; boiled leaves are used for skin issues. It is propagated by seed.

Buttonwood (*Conocarpus erectus*) (button mangrove, mangle botoncillo, mangle acemose, mang nwa): Naturally as a tree of the upper edge of mangrove swamps, buttonwood can be an exceptionally important wildlife plant in drier situations in restorations, parks, and gardens. Widespread through the Caribbean, it reaches up to 3–12 m or even 20 m but is prunable to a hedge. Attractive, rough bark provides a substrate for bromeliads and orchids, while its convoluted branches, leaves, and fruit clusters support armies of insects, which attract insect-seeking birds. Tiny, fuzzy, green flowers attract bees and butterflies. Seed-eating birds, especially bullfinches, pick seeds from 1-cm fruit. Two natural varieties, both in cultivation, have green or silver leaves. Bark has been used for tanning, and its hard wood is used for firewood and charcoal. Leaves are boiled for treating joint problems, sores, vomit-inducement, and diarrhea; and extracts have been proven to have antibacterial properties. Buttonwood is propagated by cuttings and by transplanting seedlings.

Buttonwood and Greater Antillean bullfinch.

Black olive (*Terminalia/Bucida buceras*) (Antigua whitewood, bullet tree, gregorywood, gri-gri, gué-gué, Inagua oak, oxhorn bucida, shady lady): Widespread in the Caribbean, this historically widely planted native tree is generally about 7–10 m but can reach 30 m or be pruned smaller. Straight, thick, attractively furrowed trunk and branches support insects that attract insect-eating birds. Its thick canopy of terminal blue-green leaves provides excellent shelter, very secure nest sites, and also height for perches, often used by kingbirds. Female trees produce 1/2-cm fruit in summer that are eaten by hard-billed birds such as bullfinches, parakeets, and parrots. Landscapers tend to use less-messy male trees, but these lack bird fruit, so they are not as valuable for wildlife. It is propagated by seed planted soon after falling from the tree, cutting, and air layering. It may be available in nurseries; but horticulturally, it has unfortunately become something of an old-time, out-of-favor plant.

Stoppers (*Eugenia*): Over 200 eugenia species and relatives occur in the Caribbean, over 90% being endemic. Other genera too are called stoppers. Such locally indigenous species are important candidates for native plantings as they are superior bird and wildlife plants. They are shrubs to small trees or can be pruned into short shrubs and hedges. Mixed stopper plantings provide nectar, small fruit, and invertebrates for much of the year. Dense branch structure and leaves create cover that hosts diverse invertebrates, especially spiders. Small white flowers attract bees and butterflies. Depending on species, fruit ranging from 1 mm to 1 cm are consumed by fruit-eating birds. Among well-known species widespread in the Caribbean are white stopper (*Eugenia axillaris*) (skunkwood, Spanish stopper, stopper bush); Spanish stopper (*Eugenia foetida*) (birigí, bois myrte, boxleaf stopper, guairaje, ironwood, myositis, red stopper, stopper bush, tufais); redberry stopper (*Eugenia*

White stopper.

confusa) (ironwood); twinberry stopper (*Myrcianthes fragrans*) (bois d'ine, bois mulâtre, malaguette, nakedwood, Simpson's stopper); Bahama stopper (*Mosiera/Psidium longipes*) (long-stalked stopper, sweet Margaret, wild guava). Stoppers are traditionally used for any number of ailments, including "stopping" diarrhea. Straight branches also have many uses (one name is rodwood). Nonnative Surinam cherry (*Eugenia uniflora*) is invasive into native habitats. Stoppers are propagated by seed grown in shade and by cuttings. Some species are increasingly available in plant nurseries.

Spicewood (*Myrcia/Calyptranthes pallens/Myrcia chytraculia* var. *pauciflora*) (pale lid-flower): Although spicewood is the best known, 128 species of calyptranthes occur in the Caribbean, nearly all endemic. They are superior wildlife plants. Spicewood, occurring Caribbean-wide, has multiple trunks, providing cover and structure. Fragrant clusters of white flowers lure bees and butterflies. Abundant, 1/2-cm, orange-turning-dark-purple fruit are eaten by fruit-seeking birds and by people. While tolerating a range of well-drained sites, spicewood prefers organic soil. Given the endemism of calyptranthes species, local knowledge can identify local species for use in restorations and natural gardening. They are propagated by cleaned, scarified seed. Spicewood may be available in plant nurseries.

Guava (*Psidium guajava*) (goavier, gouyave, guayaba): Considered a Caribbean native, this fruit tree was likely brought to the islands by Indigenous People. Although common in gardens, it also is a hardy pioneer species in both natural habitats and disturbed areas. It is an attractive, spreading, small tree growing up to 7 m tall and prunable, although pruning should be avoided when it is in fruit. It is a host plant for skippers, blue butterflies, and several moths. Flowers, although self-fertile, attract bees and butterflies. Depending on variety, fruit may be 2 to 10 cm in size. Larger fruit are opened first by large-billed birds then eaten by other birds. Smaller ripe fruit can be eaten directly by bats, birds, and iguanas. Once on the ground, they are attacked by invertebrates, iguanas, and mammals if present. Guava's cultural connections run deep and broad, the fruit being eaten raw, stewed, sauced, preserved, jellied, drunk, or cooked; guava duff is nearly a Bahamian national food. In folk medicine, it is used for multiple issues, and leaf extracts have been proven to be antibacterial. Propagation is by seed (with fruit quality not being assured), air layering, grafting onto stock, or root or shoot cuttings. Varieties are available in plant nurseries.

Miconias (*Miconia*) (camasey): Miconias are highly diverse in the Caribbean with 125 species, 95 being endemic. They are hardy invaders of disturbed locations. Miconia have attractive tri-veined leaves and clusters of white flowers that attract a variety of insects. Super-abundant, 3 to 5-mm, light-green-turning-violet to black fruit is available nearly year-round to fruit-eating birds, which spread seed widely. Miconias grow in shade but flower and fruit better in partial sun. Representative native species include *Miconia impetiolaris, Miconia prasina, Miconia racemosa*, and *Miconia tetrandra*. Nonnative Mexican velvet tree (*Miconia calvescens*), originally from Mexico to South America, is one of world's most invasive species, referred to as the purple plague. Nonnative varieties are available in plant nurseries; but indigenous and endemic species are better used in both natural and decorative gardens. Miconias are propagated by seed, which have a long life-span.

West Indian lilac (*Tetrazygia bicolor*) (clover ash, cordobanes, false wild guava, tetrazygia, wild guava): Other native tetrazygia species have similar excellent wildlife qualities.

West Indian lilac.

Native to The Bahamas and Greater Antilles, West Indian lilac is an understory and edge plant, although it can become a tree up to 5–10 m tall. It does exceptionally well in mixed-tree settings as its thin branches wind through available space enhancing canopy structure. Attractive and fragrant (as its name suggests), clusters of spring-summer flowers attract bananaquits, hummingbirds, and bees. They are a treat in any planting. Tiny, 1/3-cm, black fruit persist on the plant and so are available nearly year-round to fruit-eating birds including bullfinches, doves, finches, orangequits, tanagers, thrashers, thrushes, trogons, vireos, and some warblers. Leaves, fruit, and flowers are used as tonic for coughs and night sweats. Tetrazygias are propagated primarily by seed. West Indian lilac is such a beautiful plant with fragrant flowers that it is now popular in horticulture and available from some nurseries.

Snakebush (*Picramnia pentandra*) (bitterbush, bitterroot, bois poison, kafe mawon, doctor bar, ojode de peje, snakeroot, valliant garcon): Occurring Caribbean-wide, this is an attractive, shade-tolerant shrub or small tree growing up to 7 m. Its foliage supports numerous insects including sulphur butterfly caterpillars. In spring and summer, tiny, white, terminally clustered flowers attract nectar- and pollen-feeding insects. Small, 1-cm fruit persist on branches of female plants through the winter. They start green, turn scarlet, and finally black when they are eaten by species such as bullfinches, mockingbirds, and woodpeckers. As expected from its alternative names, snakebush is used in traditional medicine as a leaf tea to treat fevers including malaria, stimulate appetite, and induce vomiting. Snakebush is a hardy pioneer of coastal scrub, pond edges, and old pasture and so is well used as part of a mixed-tree or shrub planting or a diverse hedge. It is propagated by seed and may be available in specialized plant nurseries.

Gumbo limbo (*Bursera simaruba*) (almácigo, birch, birch gum, bois de' encens, chaca, chiboue, copperwood, cumalamy, gammalamme, gommier rouge, gommier wouj, gomye, gumalamy, gum bark, gum elemi, gum tree, incense tree, kalami, mastic tree, red birch, turpentine tree, West Indian birch and made-up names such as tourist tree for its skin-like, peeling red bark): Having many names suggests the plant's intimate relationship with Caribbean traditions. Gumbo limbo is one of the Caribbean's most important wildlife plants and should be in every native-species planting from restorations to gardens and also on its own streetside and in active parks. A self-seeding, pioneer tree; it grows extraordinarily fast, having a thick, brown-red-peeling trunk and spreading crown to 17 m tall. It stores water in its trunk in wet season, is deciduous in dry season, and photosynthesizes through its thin bark. Its sap repels insects but nonetheless the tree always supports impressive numbers of bees, beetles, purplewing butterflies and moth caterpillars, scale and white flies that lead to sooty mold, and other insects, all of which are hunted by birds and lizards. Tiny, green, spring flowers attract nectar-seeking insects, particularly bees. Gummy, high-energy, 1-cm fruit appears at the end of the dry season and are heavily used by resident and soon to migrate birds. It provides one of the Caribbean's most critical bird fruits. Its limbs can sprout if planted, and so it is used as a living fence. This trait also makes it resilient to storms, as its soft, brittle limbs readily break off. It has many uses including incense, insect repellent, carving, as a resinous firewood, in treating inflammation and poisonwood dermatitis, and many others. Gumbo limbo can be propagated by seed, transplanting seedlings, cuttings, and planting branches. It is available in nurseries.

Gaubán (*Trichilia hirta*) (broomstick, cabo de hacha, guaita, jobillo, jubabán, red cedar, retamo, wild mahogany): It is a widely occurring tree through the region in limestone areas from low to mid-elevation; it is a pioneer species often becoming a hedgerow component. It has scaly, brown-to-gray bark and a rounded crown of 3–15 m. It is highly attractive to wildlife. Green-white flowers attract bees, including honey and stingless bees, butterflies, and hummingbirds. Its 2-cm, red-purple, multi-chambered fruit encapsulate red-aril-covered seeds that are eaten by many species including catbirds, flycatchers, kingbirds, mockingbirds, and trogons, which disperse seeds widely. It can flower and fruit for much of the year. It has many uses throughout its range including for handles and furniture, ulcer baths, seed-oil, as a purgative, and for colds, asthma, and diarrhea. It is planted as a shade, ornamental, or street-side tree, and so it is a hardy addition to a natural planting or it is valuable on its own. It can be propagated from unprocessed, very fresh seeds with aril attached.

Mango (*Mangifera indica*) (ji mango, le manguier, mangue, pye bwa mango): There is nothing that characterizes a Caribbean community, park, or garden more than a mango tree. They are neighborhood favorites. This nonnative fruit tree is as friendly to birds and insects as it is to people. It can get quite large, 10–15 m tall, but can be kept trimmed to fit any size yard; and short-growing varieties are available. As might be expected for a widely traded commercial species, the tree hosts lots of insects; birds hunt there all the time. Small white flowers in loose panicles attract ants, bats, honeybees, beetles, butterflies, flies, and wasps. Many birds, bats, and iguanas eat young mangos, and large birds, such as parrots and woodpeckers, can break into larger fruit, after which they are open season for any fruit-eating bird and sugar-seeking insects. Mangos are best propagated by air layering and grafting to assure rapid production and fruit quality.

Gumbo limbo in fruit and Cuban parrot.

Hog plum (*Spondias mombin*) (amra, caja fruit, coolie plum, gully plum, gwo momben, jobo, mombin, Spanish plum, taperebá, yellow mombin): Hog plum, or yellow mombin, occurs naturally in the Greater and Lesser Antilles but has been introduced and naturalized around the tropical world. It can be a substantial tree up to 20 m tall with wide-spreading branches, hardy, drought-tolerant, fast-growing, and brittle. Fragrant, small, white, spring flowers attract bees and other insects. Large, 4-cm, smooth, oily, sweet to acidic, summer-fall fruit is picked at by larger fruit-eating birds and is edible being made into syrup, drinks, and jelly. Medicinally hog plum is used for almost every ailment. It is planted as a pioneer tree in forest restorations, as a shade tree, and in natural gardens. It is easily propagated by seed, but cuttings, air layering, and grafting assure fruit quality.

Palo de caja (*Allophylus cominia*) (caja, Christmasberry, kan lool, parida, rompe caldero, rois cotes, three-finger, trois feuilles, trois paroles, yanilla): Occurring in The Bahamas and Greater Antilles, palo de caja is a slender shrub to small tree, 3–7 m tall, with distinctive, triple, hairy, veined leaves. Small, white, late-summer flowers attract a variety of insects, especially bees. This lychee relative has clusters of small fruit that start yellow-turning-red just when winter resident birds need them. It is a popular medical plant in Cuba for treating diabetes, a use supported scientifically. It is propagated by seed.

Citrus (Rutaceae): Citrus, including cultivated citrus and native relatives, are superior wildlife plants. They host numerous invertebrates such as ants, aphids, fruit flies, scale, snails, spiders, wasps, weevils, whiteflies, and moth and butterfly caterpillars including the distinctive orangedog caterpillars of swallowtail butterflies. Insectivorous birds and tree lizards follow. Most flower in spring in the dry season. Fragrant flowers attract insects, especially bees and wasps. Fruit attracts fruit-eating birds, depending on size and toughness of the fruit. Domesticated citrus have wildlife values similar to native species. These originated in Asia and are extensively hybridized. Calamondins, grapefruit, lemons, Mexican limes, and oranges are commonly planted in Caribbean gardens. Jamaica, Puerto Rico, Guadeloupe, Dominica, and Martinique produce commercial exports. Citrus can be propagated by seed, but commercial varieties are best propagated by cuttings, air layering, and grafting. Cultivated varieties and some native species are available in plant nurseries, but they are under tight restrictions in some locations to control disease.

As cultivated citrus become rarer owing to disease, native citrus become even more important for wildlife. White torch (*Amyris elemifera*) (bois chandelle, candlewood, cuabilla, sea amyris, sea torchwood, smooth torchwood) occurs region-wide in coastal forest, rocky scrub, and forest edges. It is a tree, 3–10 m tall, with a slender trunk and erect branches. Its tiny, 1-mm fruit are available in winter into spring. Historically, a torch plant and its leaf tea, alone or in a blend, were used for flu. It has proven antibacterial and antifungal properties.

Yellowwoods (*Zanthoxylum*): These native citrus relatives are thorny small trees that provide protection for lizards and bird nests. Tiny, fragrant, summer-fall flowers attract pollinators. The fruit split open revealing small, hard, black seeds surrounded by a white pulp that birds eat. Yellow sandalwood (*Zanthoxylum flavum*) (misimieu, noyer, satinwood, tembetaria, West Indian satinwood, yellow sanders) was a historically important lumber tree for its durable wood. Wild lime (*Zanthoxylum fagara*) (lime prickly ash, pine jòn, pine juane) is a variably sized small to large tree that has wide-spreading, thorny limbs that thread their way through other plants providing protection. Other indigenous species also occur and should be considered for native plantings on their island.

Paradise trees (*Simarouba*): Paradise trees are fast-growing, scrub forest trees with distinctive pinnate leaves and loose spreading crown that provide cover and perch sites. They germinate and persist within canopy shade where the fast-growing, straight trunk soon can emerge from the short scrub canopy. They can readily reach 9 m tall, but they can be even taller. They make exceptional plants for restorations and for public and natural gardens. Several species occur in the Caribbean, all with similar wildlife values. Paradise tree (*Simarouba glauca*) (aceitillo, aceituno, bitter damsel, bitterwood, bois amer, bois blanc, bois frêne, dysentery bark, fwenn, matchwood, oil tree, palo amargo, paradise plum, tree of heaven) occurs in The Bahamas and Greater Antilles. *Simarouba berteroana* and *Simarouba laevis* are Caribbean endemics as is aceitillo cimarron (*Simarouba tulae*) (aceitillo falso, bitterwood, paradise tree) of Puerto Rico. Small yellow flowers attract bees. In spring, copiously produced, fleshy, red-turning-black, edible (in a pinch), 2-cm fruit have a white pulp that attracts flocks of birds even before being ripe. Simarouba have a long folk-medicine history of use for treating bleeding, dysentery, fevers, and malaria. A bitter tonic, called Jamaica bark, is taken for these and other conditions. The wood is termite-resistant, and seeds can be rendered into oil. Simarouba are propagated by dried, cleaned seed sown in the shade, cuttings, grafting, and – most efficiently – transplanting from the carpet of seedlings under mother trees. They can be found in some plant nurseries.

Muskwood (*Guarea/Caesarea guidonia*) (ball-wood, bois balé, carrapeta, cedro macho, cocillana, guaraguao, guarea, jitó, macaqueiro, redwood, requia, wild avocado): Found in the Greater and Lesser Antilles, this valuable tree, potentially up to 8 m tall, has a

Yellow sandalwood.

thick straight trunk and a wide dense canopy. It can flower year-round, but mostly in spring-summer, when its white flowers attract bees, butterflies, and hummingbirds. Clustered, 4-cm, capsule fruit attracts birds and iguanas that consume its orange-red pulp. It is heavily harvested to treat any number of ailments and for its wood used diversely in carpentry. Oil also is obtained from the wood. It is an important tree for restorations and public gardens, less so for private gardens due to its size. It is propagated by shade-grown seed.

Hibiscus (Malavaceae): Hibiscus relatives are a diverse lot of over 4,000 species, some Caribbean natives. China rose (*Hibiscus rosa-sinensis*), originally from the Pacific islands and a symbol of the tropics, exemplifies garden-worthy tropical flowers. Hibiscus flowers are recognizable from their fused, tube-like stamens emerging from often colorful petals. Another nonnative, originally African, hibiscus commonly planted in the Caribbean is sorrel (*Hibiscus sabdariffa*) (flor de Jamaica, groseille-pays, gwozey-péi, roselle), used as a tea especially at Christmas. Most hibiscus flowers produce copious nectar (enough for children to suck out) and attract their primary pollinators, hummingbirds, as well as other nectar-seekers including always-present ants, bananaquits, bats, bees, butterflies, some warblers, and wasps. Plants of the genera *Hibiscus, Pavonia*, *Sida*, and *Talipariti* are indigenous in the Caribbean.

Swampbushes (*Pavonia*): Swampbushes are among the most effective hummingbird- and butterfly-attracting plants in the Caribbean. Although the green and yellow flowers are not very obvious, swampbushes are magnets for bananaquits, butterflies, and other nectar-seeking insects, hummingbirds, and warblers. Swampbushes can come alive with nectar-feeders and their avian and lizard predators. Thick shrubs to small trees growing to

Bahama swampbush.

about 4 m with spindly, brittle branches, they can produce mass clumps that offer superior cover. Bahama swampbush (*Pavonia bahamensis*) (Bahama marshbush) is endemic to the Lucayan Archipelago. Mangrove mallow (*Pavonia paludicola*) is Caribbean-wide. Like other hibiscus flowers, swampbushes are easily propagated by cuttings.

Mahoes (*Hibiscus/Talipariti, Thespesia*) (cork trees, tree hibiscus): Mahoes are variably large (3–10 m), spreading trees nearly ubiquitous in sheltered coves along the Caribbean coast. They also occur in fresh water and can be planted on drier sites. Cork tree canopies provide excellent structure for birds, crabs, iguanas, snails, and other animals along protected coasts. The 4–8-cm flowers attract bees, their primary pollinator, other nectar-seeking insects and hummingbirds. Seaside mahoe (*Hibiscus tiliaceus/pernambucensis/Talipariti tiliaceum*) (beach hibiscus, coast cottonwood, cotton tree, mahaut, mahoe, mahot, majagua, mausa, purau, sea hibiscus, tree hibiscus) is pantropical, but may be a Caribbean native having dispersed there long ago. It has yellow flowers with red centers. Blue mahoe (*Hibiscus elatus/Talipariti elatum*) (mahaut, majagua azul, majó azul) occurs in the Greater and Lesser Antilles and is Jamaica's national tree. Mahoe (*Thespesia populnea*) (Indian tulip tree, milo, Pacific rosewood, portia tree) is from the Pacific but disputedly may also be a New World native. It has yellow flowers. Hardy and adaptable, cork trees are valuable in forest and coastal restoration. They are propagated by seed in the sun. They also self-propagate by sprouting prostrate stems and so can be grown from cuttings even from large branches.

Fanpetals (*Sida*): Smallish hibiscus relatives, these are annual and perennial shrubs of about 2 m tall. Fanpetals are host to skipper butterfly caterpillars. Although self-pollinating, flowers attract bees, butterflies, and wasps. A number of species are native to the Caribbean, including island endemics, but it is not clear even to taxonomic specialists how many species of *Sida* there are. This is not so important if plants native to the island are used in plantings. Broomweed (*Sida acuta*) (bali midi, baslier savane, common wireweed, wireweed), a Caribbean-wide species, is a straggling shrub with small hibiscus-like flowers. Being very hardy, it is an aggressive pest elsewhere in world. Other Caribbean-wide fanpetals include bracted fanpetal (*Sida ciliaris*) and common fanpetal (*Sida ulmifolia*) (common wireweed). Being weedy and pollinator-friendly, it is beneficial in a natural garden. Fanpetals are propagated by seed. Local species, perhaps endemic, may volunteer.

Ceiba (*Ceiba pentandra*) (fromager, giant kapok, la ceiba común, mapou coton, silk-cotton tree): Ceiba and related Old World kapok trees (*Bombax ceiba*) have similar wildlife values. They are huge trees, one of largest in the world, towering to 70 m with a trunk 3 m or more wide plus impressive buttresses. The crown is broad, spreading upwards of 60 m. It is a dry-season deciduous tree. Pink flowers cover the canopy before leaf-out. Flowers are primarily bat-pollinated, but nectar and pollen also attract bananaquits, hummingbirds, and other birds, and lots of insects. Flowers fill with water and become a water source for birds and insects. Nectar, sugary water, flowers, and insects are all important in the dry season for many birds including blackbirds, grassquits, hummingbirds, orioles, parakeets, parrots, tanagers, warblers, and woodpeckers. A tree produces hundreds of 15-cm seed pods filled with fluffy fiber, which traditionally was used for stuffing. Starting with Taíno, ceibas are historically important cultural elements and protected by communities throughout the region and beyond. They are considered sacred in many traditions, including as duppy trees. Ceibas are native to dry-season forests and so are hardy and require no care. Both ceiba and kapok trees are long-lived and have been proudly planted in town squares, public parks, and botanical gardens. Surprisingly, they can also fit into private gardens, although as dominating specimens. They need a dry season without irrigation. Propagation is by seed and by cuttings from actively growing branches.

Capers (*Quadrella/Capparis/Cynophalla*): Over 20 caper species occur in the Caribbean, all with wildlife and natural garden values. Worldwide, caper fruit and buds are used as food and similarly are superior wildlife plants, highly suitable for restoration projects, public parks, and private natural gardens. Jamaica caper (*Capparis/Quadrella cynophallophora/jamaicensis*) (black caper, black willow, bwa dajan, bwa kaka, caper, headache bush, wild orange), occurring Caribbean-wide, develops a densely branched crown of attractive shiny green leaves, providing excellent cover. It is usually a shrub but potentially a tree growing to 5 m and prunable to thick, attractive wildlife-friendly hedge. In spring and early summer, fragrant, attractively filamentous, white-turning-violet flowers entice numerous butterflies and other pollinators. In summer, notable 30-cm pods break open revealing black seeds surrounded by a red pulp consumed by birds. Traditionally used in folk medicine, its leaf tea is used for headaches. Jamaica caper is tolerant of salt and drought and can grow almost anywhere. Limber caper (*Capparis/Quadrella/Cynophalla flexuosa*) (bayleaf capertree, bwa kaka, dog caper), occurring Caribbean-wide, is a scrambling, woody, viny shrub. Other indigenous species also might be considered for natural gardens and other plantings. Capers are readily propagated by seed and by transplanting sprouts. Popular species are available at some plant nurseries.

Papaya (*Carica papaya*) (fruta bomba, lechosa, lechoza, mamón, olocotón, papaw, papay, papayo, papayón, pawpaw, tee melon): Not only one of the more useful all-round

Jamaica caper and bananaquit.

Caribbean garden plants, papayas are also excellent plants for insects, birds, and other wildlife. Considered native, having been introduced by Indigenous Peoples, they are weedy, hardy plants to be found throughout the region from imbedded in tropical forests to waste places. Most papaya plants are tree-sized, growing quickly to 7 m although individuals and varieties may be shorter. A papaya plant is short-lived with a soft, hollow, straight trunk and a succession of umbrella leaves. As both a native and commercial species, it not-unexpectedly supports many insects such as ants, aphids, grasshoppers, moths including sphinx moths and their hornworm caterpillars, spider mites, and white flies. The funnel-shaped whitish flowers attract numerous potential pollinators including honeybees, butterflies, hoverflies, wasps, and, at night, moths, their primary pollinator. Small birds hunt insects along the large leaves, stalks, and flower and fruit clusters. Both male and female or bisexual plants are needed for fruiting. Fruit can appear any time of the year, ranging from a few cm to 20 cm in size, depending on the individual plant and variety. The fruit attracts aphids, butterflies, flies including papaya fruit flies and white flies, moths, spider mites, wasps, as well as bats, iguanas, and fruit-eating birds. Tough-billed birds such as woodpeckers and parrots cut into the skin and, once opened or on the ground, the fruit is fair game for any fruit-seeking bird, reptile, or crab. Usually, the plant is so prolific that there is enough fruit to share with birds; or fruit can be bagged. Papayas can be grown from seeds that are first dried and cleaned. But seeds sprout best after being eaten and deposited by birds and so may volunteer, but volunteer seedlings need to be transplanted before the taproot is too large. They can also be started from side shoots and cuttings using rooting media. There are many named cultivars available in plant nurseries and at market.

Coccolobas (*Coccoloba*): Over a hundred species of coccolobas are native to the American tropics. They are single- or multi-stemmed, hardy, fast-growing shrubs, trees, or vines, resistant to wind, salt, sand, and dry seasons. The thick (and in some species very large) leaves and peeling bark always seem to attract a dense community of insects including ants, beetles, bugs, caterpillars, flies, spiders, termites, and wasps, all of which are food for insect-eating birds, particularly warblers but also gnatcatchers, kingbirds, mockingbirds, vireos, and woodpeckers. The tiny, fragrant flowers attract butterflies and bees, including honeybees. Clusters of grape-like fruit are eaten by iguanas and a long list of birds, especially white-crowned and scaly-naped pigeons and also wintering catbirds, mockingbirds, thrashers, and woodpeckers, and some by people. Both male and female plants are required for fruiting.

The most recognizable and popular, likely, is seagrape (*Coccoloba uvifera*) (baygrape, coccoloba, grape, grapetree, Jamaican kino, kino, raisinier bord-de-mer, raisin la mer, rezen lamè, seaside grape, uva de playa, wild grape). Seagrape is a dominant plant on beach dunes and shores throughout the Caribbean. It can be a spreading tree 7 m or taller in forests or can be trimmed to a bush or hedge, as it is adapted to be hedged naturally by salt spray and wind. Massive kidney-shaped leaves (to 10 cm or more) provide exceptional cover; but their accumulation inhibits other plants so the ground is open but leaf-mulched. Ground-foraging birds and ground lizards search through the fallen leaves. Fruit is edible, sort of; but its pectin content makes for good jelly. The collecting

Seagrape with Hispaniolan woodpecker.

difficulty is that individual fruit within a bunch ripens independently. Its bark is traditionally boiled for diarrhea, its wood used for carpentry, and its sap for dye and tanning.

Another Caribbean-wide species is pigeon plum (*Coccoloba diversifolia*) (ciruela de paloma, dove plum, maivisse, pigeon seagrape, raisinier, tie-tongue, uva de paloma, uvilla, zamon marron): Pigeon plums tend to have straight, attractive trunks with narrow crowns, which make them good for crowded spaces and mixed-species settings. At the edge of a forest or natural garden, they become part of the border hedgerow. Pigeon plum's 1/2-cm white flowers and 1–3-cm fruit are clustered on spikes where they are readily available for birds. The fruit is edible, although palatability varies among trees and motivation of the consumer.

Other coccoloba species include Caribbean-wide whitewood seagrape (*Coccoloba krugii*) and grandleaf seagrape (*Coccoloba pubescens*) (Eve's umbrella, gamelle, largeleaf seagrape, moralón, mountain-grape, raisin grandes) from the Lesser Antilles and Puerto Rico. With so many native and endemic species, local species can be used in natural plantings. Native coccolobas are essential in restorations and hardy in parks and natural gardens, as a street tree, and in commercial settings. Coccoloba species and pigeon plum populations indigenous to an island can be sought out and used. They grow readily from seed and seedlings can be transplanted. They can grow in partial shade. Pigeon plum and seagrape are available from some plant nurseries.

Guineaweed (*Petiveria alliacea*) (anamú, apacin, ave, feuilles ave poules, garlicweed, grannybush, guinea bush, Guineahen bush, gullyroot, herbe aux petevere, koujourouk, mapurite, mucura, obeah bush, pondbush, stinkweed, tipi, vèvènn): Occurring Caribbean-wide, a well-known and well-appreciated, short, upright, 1-m shrub with small leaves and deep-running, garlicky-smelling roots. It is a hardy plant of rocky substrate and vacant lots. In winter and spring, tiny white flowers on long spikes attract bees. Bullfinches and seed-eating birds eat the dry fruits. Guineaweed is used in bush medicine for just about everything – pain, skin, colds, sores, cuts, and as an aphrodisiac (with rum, of course). Its name of granny bush is a testament to traditional use in childbirth. In some places, it is available in powder or pill form as anamu. Like all plants named obeah, it is used for contending with evil eye and ghosts or ailments they might cause. Guineaweed is propagated by cuttings kept damp. Because of its deep roots, it does not transplant well.

Bloodberry (*Rivina humilis*) (baby peppers, coralberry, corralito, inkberry, pigeonberry, rougeplant): Occurring Caribbean-wide, this common, viny shrub about 1 m tall is found broadly along the edges of forest, scrub, roadsides, and disturbed sites. It is a drought-resistant, weedy, shade tolerant, perennial – perfect for edges of a natural area. Sprays of attractive small white or pinkish flowers attract bees. Small, 5-mm, bright-red, year-round fruit attract small fruit-eating birds. It is propagated by seed. It has been exported around the tropical world, and, once established, it will readily self-seed.

Bloodberry and palm warbler.

Bougainvillea in flower.

Bougainvillea (*Bougainvillea spectabilis*) (bugambilia, buganvilia, purple bougainvillea, trinitaria veranera): Possibly being the most recognized tropical plant, bougainvilleas are shrubs or vines. They are nonnative but widely planted throughout the Caribbean and a hallmark of photogenic tropicalness. Originally from eastern Brazil, it has significant wildlife values. On its own, in a butterfly and hummingbird garden, invading a tree, covering a fence, or climbing an urban building, bougainvillea provides excellent cover for birds and other wildlife within its thorny branches. It is prime real estate for nests. Insects and birds are attracted to flowers. There are many horticultural varieties and flower colors, but wildlife benefits from plants with large, scrambling growth form, fully equipped with painful thorns. Despite their friendliness to bare feet, thornless varieties do not provide as much security and tend to be small. A good dry season is needed for bougainvilleas to bloom profusely. They are best propagated by cuttings and are readily available in plant nurseries.

Blollies (*Guapira*): At least 15 blolly species are known from the Caribbean, some historically harvested as beefwood, and so are now rare. They are attractive shrubs to small trees of 3–7 m tall, narrow with erect, sometimes drooping branch tips. They flower and fruit for much of the year. Flowers attract bees, moths, other insects; and the small fruit with a distinctive indentation are edible and sought after by fruit-eating birds especially pigeons and thick-billed vireos but also iguanas, mockingbirds, spindalis, tanagers, and warblers. All serve well in naturalistic garden settings. Small-leaf blolly (*Guapira discolor*) (beeftree, beefwood, blolly, long-leaf blolly, narrow leaf blolly, pigeonberry) is Caribbean-wide and now protected in some jurisdictions. Broadleaf blolly (*Guapira obtusata*) (beef tree, beef, blolly, bloodberry, bloodwood) occurs in The Bahamas and Greater Antilles. Black loblolly (*Guapira fragrans*) (black mampoo, mapau) occurs through the Greater and Lesser Antilles. Blolly wood is used for numerous purposes requiring strength. They are propagated by seed germinated in the shade.

Golden-spined cactus and Venezuelan troupial in Bonaire.

Cactuses (Cactaceae): Cactuses are excellent wildlife plants. Their spiny modified leaves provide superior security and protected spaces for a wide variety of animals. Hummingbirds, cavity-nesters and other birds take advantage of them as nest sites. Yellow-shouldered amazons chew them for water. Depending on species, beautiful flowers attract bananaquits, bats, bees, butterflies, hawkmoths, hummingbirds,

Infrou prickly pear cactus and bird nest.

orangequits, orioles, and others. Small fruit are consumed by fruit- and seed-eating birds, whereas larger fruit are eaten by larger birds and iguanas. Once open, fruit may be eaten by smaller birds and other animals. Cactus, of course, is ideal for xeriscaping but often overlooked more generally, except where they are used traditionally as living fences. The Caribbean has a wide selection of native cacti, some occurring region-wide; others are island endemics, which should be sought out for use in natural plantings.

Among natives, prickly pear species (*Opuntia*) (fig, nopal, raquette volante, sabra, tuna) are widespread, easily established, and highly adaptable to gardens. Necklace prickly pear (*Consolea/ Opuntia moniliformis*) (alpargata, patte-de-tortue, patte tortue, pat tôti, raquette espagnole) is a Caribbean-wide, columnar or tree cactus, 3–4 m tall. It occurs naturally in dry limestone forest, is in cultivation, and used in gardens. More

Infrou prickly pear cactus in flower.

range-restricted, infrou (*Opuntia caracassana/wentiana*) (common prickly pear, spaanse juffer, tuna) is from the Leeward Antilles. Turk's cap (*Melocactus*) (barrel cactus, melon cactus, pincushion, pope's head, tête anglaise) is widespread with 15 species in the Caribbean and 9 in Cuba alone. The Turks Islands are named after it. A plant of rocky areas and rocky shores, it bears a distinctive red, wooly, bristly cap and has an edible fruit. Melocactus make excellent garden plants; some are cultivated. Cereus (*Hylocereus*) is a vining night-blooming cactus from the Greater and Lesser Antilles. Night-blooming cactus flowers attract bats, their principal pollinator. Dildo cactus (*Pilosocereus polygonus*) (Deering's tree cactus, Key tree cactus, tree cactus) from the Greater Antilles is a widespread columnar or tree-form reaching 9 m. The endemic columnar golden-spined cactus (*Pilosocereus lanuginosus*) and dagger cactus (*Stenocereus griseus*) (pitaya and pitayo de mayo), along with Turk's cap and prickly pear, dominate the distinctive cactus scrub habitat of the Leeward Antilles. These columnar cactuses can reach heights of 6 m and are used as living fences.

Several nonnative cactus species are common. Cochineal nopal cactus (*Opuntia cochenillifera*), originally from Mexico and Mesoamerica, is historically significant as a substrate for cochineal scale insects (*Dactylopius*). The red dye produced when squashed was once an extremely valuable export. Indian fig cactus (*Opuntia ficus-indica*), also originally Mexican, is the edible commercial species. Erect prickly pear (*Opuntia stricta/dillenii*) is a widely dispersed Mexican native, also edible, but has become a massively serious invasive in Australia and Africa. Originally Mexican but now widely cultivated pitahaya (*Selenicereus*) (dragonfruit) provides both fruit and wildlife structure. Cactus species out of range are often hard to identify by their vegetative growth forms alone. In addition to commercial applications, cactus has always been culturally important for food, tonics, soap, and skin products. They are propagated by seed, planting stem segments, and division of pups. There are differences, melocactus is propagated by seed but opuntia are readily started from pads. Specimens are available in plant nurseries but likely not local species.

Wild dilly (*Manilkara jaimiqui/bahamensis/emarginata*): Native to The Bahamas, and Greater Antilles, dillies are usually shrubs but can be small trees of 3–5 m tall. Many birds forage in the branches and tight clusters of leaves for a wealth of insects, spiders, and other invertebrates that inhabit the foliage. Spring-summer flowers attract bananaquits and hummingbirds. Being relatives of commercial sapodilla (*Manilkara zapota*) (chicle, dilly tree, naseberry, sapote), their smaller 2-cm fruit are more bird-friendly than the larger slow-ripening sapodilla and are consumed by many fruit-eating birds as well as by bats and iguanas. Naturally found in rocky shoreline scrub, it does well in coastal, natural, and dry gardens. Wild dilly is readily propagated by seed and can be transplanted.

Satinleaf (*Chrysophyllum oliviforme*) (caimite, caimitillo, damson plum, kaymit mawon, olive plum, saffron): Occurring Caribbean-wide, satinleaf is a strikingly beautiful tree of 3–10 m tall, with attractive, smooth, and grayish bark. The noticeable 8-cm leaves are glossy green on top and fuzzy brown to gold below. It attracts many insects and birds, especially wintering birds, which forage in the canopy. Tiny, 1-mm, fragrant, greenish-yellow, fall-winter flowers attract bees and many other insects. Sweet, 1-cm, purple, late winter-spring fruit attract larger fruit-eating birds. They are edible but chewy. It can be a specimen tree but does well in a mixed-tree forest. Leaf tea is used for circulatory issues;

Wild dilly.

the leaves contain coumarins from which warfarin blood thinner is derived. Satinleaf is readily propagated by seed, transplanting seedlings, and may be available in plant nurseries.

Bully trees (*Sideroxylon*): Over a dozen sideroxylon species occur in the Caribbean, some of conservation concern. They have similar wildlife values. A dense canopy provides cover and structure; flowering and fruiting seasons are prolonged with both flowers and fruit being heavily used by birds and other animals. Bully trees are hardy and important options for restorations, regreening, and natural gardens within their native ranges, also as shade and park trees. They are propagated by scarified and soaked seed or by transplanting seedlings, all kept in shade. Bully species are sometimes available in plant nurseries.

Mastic (*Sideroxylon/Mastichodendron foetidissimum*) (acoma, almástiga, ascomat, bully-mastic, caguani, caya, false mastic, ironwood, jocoma, masticwood, mastwood, sabina, saguinaria, yellow mastic, wild mastic): Occurring Caribbean-wide, mastic is a potentially large, fast-growing, soil-tolerant tree, up to 20–25 m tall, with trunks up to 3 m in diameter. Inconspicuous, yellow, cheesy-smelling flowers attract flies. Its yellow-orange, 2-cm fruit, called jungle plum, is available in winter to fruit-eating birds especially pigeons. Edible although gummy and acidic, the fruit was indigenously cultivated in Mexico. It is a species of many names reflecting its historically important uses. The sap, similar to mastic gum of the Old World, was one of the colonial era's sought-after natural products. It is a tree of limestone dry forest and is drought-, salt-, and hurricane-resistant.

Milkberry (*Sideroxylon americanum/Bumelia americana*) (bo-hog, wild saffron): Occurring in The Bahamas and Greater Antilles, milkberry is a small tree, 3–7 m tall, with a short trunk, spreading crown of dense branches, providing nest sites, with attractive leaves, green above, rusty below. It can bloom and fruit multiple times a year. Fragrant, green flowers attract insects; and the small, 1-cm fruit is consumed by fruit-eating birds.

Willow bustic (*Sideroxylon salicifolium/Bumelia/Dipholis salicifolia*) (acomat marron, cassada wood, caya rubia, m'panache, white bully): Found Caribbean-wide, willow bustic is a tall, slender shrub to small 5-m tree with spreading upright branches and downward-drooping, dark-green leaves. White to green, 3-cm, grape jelly-smelling flowers, clustered on branches, attract insects; and fruit-eating birds readily take the 1-cm, sweet, black fruit. As a scrub plant, it does well planted. The tree's narrow growth form makes it suitable for tight spaces in a diverse native garden and in commercial settings.

Satinleaf.

Featherbed (*Diospyros crassinervis*) (boawood, hard bark, (less politely) stiffcock, (recently) natural viagra): Occurring in The Bahamas and Greater Antilles, this is a shrub to small 7-m tree with dark-green, stiff, potent-smelling leaves. Flowers are tiny, attracting bees. A persimmon relative, it produces 2-cm, yellow-turning-black fruit in late winter and spring, which are well appreciated by birds at the end of their wintering season. From its alternative names, it has uses in bush medicine as an essential ingredient in teas aimed at men. For people, the fruit can be eaten directly. Irrespective of other uses, it does well for birds and other wildlife in natural, herb, and medicinal gardens. It is propagated by cuttings and by transplanting.

Myrsines (*Myrsine/Rapanea*) (colicwood, rapanea): There are hundreds of species of pantropical myrsines. *Myrsine cubana*, from The Bahamas and Greater Antilles, and *Myrsine guianensis*, from the Greater and Lesser Antilles, are common, native, forest-edge shrubs or small trees, 7 m tall, having slender crowns and attractive, upward-facing, smooth, thick, leathery, dark-green leaves. Blooming in winter and spring, numerous, small, 1/2-cm, cream-color-turning-red/brown flowers attract bees and other insects. Among myrsines' interesting features is that numerous fruit are unusually situated, lining the branches of female plants, providing easy access for birds. The 1-cm, black fruit provide bird food into winter dry season. The shrubs' narrow shape is conducive to tight, mixed-species plantings, as an edge plant, attractive border screen, or as part of a diverse hedgerow. Myrsine is not specific as to soil conditions, either dry or damp. Propagation is by seeds and transplanting small plants. It, or nonnative species, is sometimes available in plant nurseries.

Ardisias (*Ardisia*): There are about 500 species of ardisias worldwide including species in horticulture, Caribbean natives, and some that are invasive pests outside their native ranges. Ardisias are shrubs or small trees with the height of 5–7 m that occur in semi-shady forest understory and edges. The 1-cm white flowers form showy terminal clusters that can fragrantly cover the shrub attracting butterflies and hordes of other nectar-seeking insects and hummingbirds. The 1-mm fruit is similarly interesting, starting green turning pink, red, and then purple, sometimes in the same cluster. It is consumed by most of the fruit-eating birds, especially through the winter, including bullfinches, catbirds, mocking-birds, vireos, warblers, and woodpeckers. Hardy and wind-resistant ardisias are useful in restorations and attractive in natural gardens, especially in diverse hedgerows where flowers, fruit, and insect hoards can be appreciated. Marlberry (*Ardisia escallonioides*) (island marlberry, marbleberry) and Guadeloupe marlberry (*Ardisia obovata*) (mamayuelo) are 2 of the more widespread native species. Native marlberry may be available in plant nurseries but nonnative ardisia, the taxonomy of which is confused at best, are more readily available and often mislabeled as the native species. These are usually coralberry (*Ardisia crenata/crenulata*) (Christmasberry) or shoebutton ardisia (*Ardisia elliptica/polycephala/solanacea*) (duck-eye ardisia), both aggressive Asian species that should not be planted. Ardisias' invasiveness is a tribute to their fruit's attractiveness to birds. Marlberries are hardy plants, especially well adapted to calcareous and sandy soils. To avoid aggressive exotics, it is best to propagate native species from an island by cuttings and seeds, although these may take patience, taking about 6 months to germinate.

Jacquinias (*Jacquinia*): The several Caribbean jacquinias are slow-growing, salt- and drought-resistant, hardy, primarily coastal scrub plants. They are shrubs to small trees growing to 4 m with attractive stiff leaves and numerous short, spreading branches, creating a mounded shape and providing cover and sturdy nest sites. During a prolonged flowering season, fragrant white flowers attract pollinating insects. The similarly attractive 1-cm fruit, turning yellow to orange-red when ripening, are available mostly in fall and winter attracting all the fruit-eating birds. Native species include Joewood (*Jacquinia keyensis*) (cudjoe wood, ironwood, Joebush, washbush); braceletwood (*Jacquinia arborea*) (barbasco, bois-bouc, jacquinier bos-bracelet, pica, pie-crust torchwood, steenhoot, torchwood, zucares); bois bouc (*Jacquinia armillaris*) (armed jacquinia, baylum, braceletwood, torchwood); and ois bande (*Jacquinia berterii/berteroi*) (berter's jacquina, bois casse cou, casse cou, coppice Joewood, olivier). Although their taxonomy has been confused through the years, the species function similarly and admirably in natural plantings within their range. They are suitable for diverse natural gardens and even commercial plantings and containers. Historically, they were torch plants and used for their hard wood. Island-specific indigenous species have the greatest conservation value. They are propagated by seed and cuttings; and Joewood and braceletwood may be available in plant nurseries.

Snowberry (*Chiococca alba*) (aceitillo, bejuco de berac, bejuco de berraco, cahinca, canica, caninara, common snowberry, consuelda, croc soris, kimak, lagrimas de Maria, lagrimas de San Pedro, madreselva, milkberry, pisabed, preja de raton, quimaque, ratroot, suelda, West Indian milkberry, West Indian snowberry): With so many names, snowberry

Guadeloupe marlberry.

Joewood.

is clearly a popular plant. Widespread through the Americas, it is generally a shrub with viny tendencies that can occupy shrub-layer gaps reaching 6 m. Tubular, fragrant, yellow flowers attract bees and butterflies. A coffee-relative, the noticeable white fruit stand out from dark-green foliage attracting fruit-eating birds. Two other indigenous chiococcas occur in the region. Roots have traditional uses as a diuretic, emetic, and laxative; and, in fact, chemicals (saponins) from the roots do have anti-inflammatory activity. It is propagated by seed and cuttings and has been cultivated for some time and so may be available in garden shops.

Lilythorns (*Catesbaea*): Seventeen native lilythorn species occur in the region, mostly in the Greater Antilles. All provide superior wildlife cover in their canopy full of wildlife-protecting thorns. Slow-growing, they can be a shrub to small tree to about 3–4 m tall. Bell-shaped flowers attract hummingbirds. A coffee-relative, edible 1–5-cm fruit remain on the plant for a long time and, when ripe, are eaten by larger birds, especially by doves, parrots, and thrashers, and by people. Species include smallflower lilythorn (*Catesbaea parviflora*) (dune lilythorn) occurring Caribbean-wide; wild guava (*Catesbaea spinosa*) (lilythorn, prickly apple, Spanish guava) from The Bahamas and Greater Antilles; and tropical lilythorn (*Catesbaea melanocarpa*) from Puerto Rico, Virgin Islands, and some Lesser Antilles. Tropical and smallflower lilythorn are of conservation concern. Other indigenous lilythorns also can add a conservation dimension to natural-area plantings within their range. Natively occupying partially shaded sites, they are useful as a hedge or within a diverse hedgerow providing a thorny barrier to dogs, cattle, and people. Lilythorns are propagated by tiny dry, crushed seeds in the shade but more assuredly by cuttings.

Blacktorch (*Erithalis fruticosa*) (jayajabico, manglillo, parrot apple, pigeonberry, tea, teillo): Widespread in the Caribbean, blacktorch is one of the region's most important bird plants, in part because its fruit is available in winter for neotropical migrants, especially wintering warblers and vireos. Although it can grow to 5 m or more, usually it is shorter. Thin branches covered with small shiny leathery leaves provide cover for small birds. Its white flowers attract bees and butterflies. A coffee-relative, its small black fruit are produced in summer and fall and are retained through the winter. Historically, its wood was prized for its hardness and durability, such as for fence posts. It also was a natural torch. Medicinally, it is used as a bush tea or bath for skin issues. Common in coastal scrub and forest edge, it is salt-tolerant and hardy and can be planted as an accent shrub at the edge of a natural garden or within a diverse hedgerow. It is propagated from seed, cuttings, and by transplanting seedlings.

Velvetseeds (*Guettarda*): Several species of velvetseeds are native to the Caribbean, all of which are important plants for birds and other wildlife owing to their structure, flowers, and fruit. They are shrubs or small trees, with the height of 2–5 m, but can be taller. Their open canopy of ascending branches provides cover. Blooming year-round but peaking in spring-summer, the tubular, fragrant, white to red, 1/2-cm flowers attract bees and butterflies. A coffee-relative, many birds eat velvetseeds' small, 1/2 to 1-cm fuzzy fruit. Among the species are smooth-leaf velvetseed (*Guettarda elliptica*) (common velvetseed, oval-leaf velvetseed), occurring Caribbean-wide; rough velvetseed (*Guettarda scabra*) (velvetberry, wild guava), also Caribbean-wide; and frogwood (*Guettarda krugii/inaguensis*) (frogbush, Kruger's velvetseed), from the Lucayan Archipelago and Puerto Rico. They do well in diverse hedgerows, edges of natural gardens, or as living fences. Velvetseeds are propagated with difficulty from seed but readily from cuttings. Rough velvetseeds may be available in plant nurseries.

Shining stenostomum (*Stenostomum lucidum/Antirhea lucida*) (aguacatillo, almorrana, amarilla, antirhea, avocat marron, bija cimarrona, bois patate, mapou noir, mutton polly, palo de cuello, palo de gallina, palo ilorón, raizu, zabocat marron): Occurring in The Bahamas, Greater Antilles, and Virgin Islands, this well-known plant is a smooth-barked shrub or large tree reaching 13 m. It is widely distributed in scrub and dry forests. Numerous, pretty, 1/2-cm, white, tubular flowers attract nectar-seeking insects. A coffee-relative, small, 1/2-cm, red-turning-black fruit are readily eaten by bananaquits, flycatchers, spindalis, and tanagers. Its leaves are used in tonics for colic. Other stenostomums also occur in Caribbean such as pegwood (*Stenostomum coriaceum*) (leathery stenostomum) from the Lesser Antilles, Jamaica, and Puerto Rico. All have similar wildlife values. They are propagated by processed seeds, cuttings, air layering, or just burying a bent branch until rooted. Shining stenostomum is available in plant nurseries.

Firebushes (*Hamelia*): Caribbean firebush species are among region's most important butterfly and hummingbird plants and essential components in any restoration or wildlife-attracting garden. Best known is firebush (*Hamelia patens*) (bálsamo, busunuvo, Doña Julia, fleur corail, hummingbird bush, koray wouj, pata de pájaro, ponasí, redhead, scarlet bush). Occurring Caribbean-wide, firebush is a shrub best treated as a potential tree reaching 6 m or more. It grows quickly and widely with branches stretching into available space putting its flowers well up into the canopy of a native planting. It also volunteers easily and so suggests where it might grow. It is a valuable addition to hedgerows and orchard plantings, adding significant wildlife value to the spaces between cultivated trees. Tall, spreading trees provide the most value to wildlife. In more controlled settings, perhaps its greatest issue is how to control its outstretching growth, but it takes severe pruning well, down to hedging. It hosts an array of caterpillars including sphinx moths and other invertebrates. Insectivorous birds, tree lizards, and spiders hunt the branches. Its main wildlife feature is its tubular, red, orange, and/or

Firebush.

yellow flowers produced nearly year-round but especially in spring and summer. Its value as a nectar source – and attractant for bananaquits, butterflies, hummingbirds, and other nectar-seeking insects – can hardly be overstated. There is enough nectar that it can be sucked from the base of a plucked flower. Small, 1/2-cm, yellow-turning-orange then red then purple-black fruit are eaten by a litany of smaller fruit-eating birds. Its great wildlife properties have made it a common plant in horticulture. *Hamelia patens* has 2 subspecies and other cultivated varieties. Nonnative, cultivated, Mexican-derived plants tend to have yellower (not redder) flowers and slick (not fuzzier) leaves in whorls of 3–4; but these characteristics are not uniform. Other species also occur in the Caribbean. Bahama firebush (*Hamelia cuprea*) from the Greater Antilles has coppery leaves and yellow flowers. Balsamillo (*Hamelia axillaris*) (guayabo negro) is a short understory plant with yellow flowers and red fruit. *Hamelia papillosa* is a Jamaica endemic. The identification of indigenous stock can be difficult. Cultivated firebush varieties tend to be not as good as wildlife plants and may hybridize with natives; and these may be what are mostly available in nurseries. The best approach is to find wild specimens that appear to be native and propagate these by cuttings or transplanting volunteers.

Indigoberry (*Randia aculeata*) (arbol de navidad, bois-lance, box briar, cabai, croca chien, inkberry, goat horn, kròk chen, palo de cotorra, pricklebush, resuele-suele, steelwood, tintillo, white indigo): Being a wild coffee-relative widespread species throughout the Caribbean, indigoberry is a shrub or small tree growing to 3 m with leaves clustered at the ends of stiff, thorny, horizontal branches, which provide cover on the outside and open space within for birds, lizards, and other small wildlife. Leaves provide habitat for many insects, including sphinx moth caterpillars. Tubular, fragrant, 1-cm, white flowers attract insects, especially butterflies. The 1/2-cm fruit, mostly available in fall and winter although in some cases year-around, attracts all the fruit-eating birds. Although edible by people, the fruit does turn tongues blue. It is a common native pioneer plant with salt and wind tolerance and so is a tough plant that does well in gardens, particularly near the coast. They can be accent shrubs, part of buffers, borders, or in mixed-species shrub or tree settings. Winter fruiting makes it a Christmas plant. Stiff branches have had many uses. Medical uses include for fever and stomach cramps. It is propagated by seed, cuttings, and transplanting and is available in plant nurseries.

Exostemas (*Exostema*): Well diversified in the Caribbean, there are 24 species in the Greater Antilles alone. Inkwood (*Exostema caribaeum*) (butterbough, Caribbean princewood, Caribee bark, chandelle anglaise, Jesuit bark, prince's torch, quinquina pays) occurs throughout and beyond the Caribbean. It is a hardy, adaptable, slender shrub to an 8 m tree with distinctively erect, slender, spreading branches but can be kept pruned. Large, fragrant, extremely showy, white, spring-summer flowers at branch tips attract bees, butterflies, hummingbirds, and moths. A coffee-relative, its 1-cm, purple-black fruit attract fruit-seeking birds. Inkwood is a plant of dry, rocky, and sometimes arid soils from coastal scrub, to pinelands, to high elevations. As its large range suggests, it is hardy and, so, is adaptable to natural plantings. Twigs and bark steeped in gin are used as bitters for very nearly any ailment. Other exostemas are more range- and habitat-restricted. Jamaican bark (*Exostema sanctae-luciae*), a Lesser Antillean endemic, was discovered to be a quinine substitute in the 1800s and is still an important tonic ingredient in Saint Lucia. All exostemas are valuable wildlife plants and should be considered for planting within their ranges. They are propagated easily by transplanting seedlings. Because of its showiness, inkwood is available in plant nurseries.

Beach creeper (*Ernodea littoralis*) (coughbush, ernodea, golden creeper): Occurring throughout the Caribbean, beach creeper is a viny shrub of beach, scrub, pineland, and disturbed sites. Small, 2-cm green to yellow leaves clustered at the end of the thin prostrate stems form mats 1 m tall providing cover. Blooming year-round, tubular, red, or white flowers attract bananaquits, butterflies, and hummingbirds. Being a coffee relative, its golden berries with tiny, 1-mm, black seeds are eaten by small seed-eating birds, especially bullfinches and grassquits. As a beach plant, it is tolerant of dry conditions and neglect and is good for beach ecosystem restoration, xeriscaping, borders, as well as for butterfly, rock, and natural gardens. It does not take overwatering. Tea from shoots is used to treat coughs. At least 9 other ernodea species occur in the Caribbean having similar values. They are propagated by seed, covering a stem with dirt until rooted, or transplanting seedlings. Beach creeper may be available in plant nurseries.

Wild coffees (*Psychotria*): Wild coffees are excellent – actually essential – plants for ecosystem restorations, parks, nature centers, and natural gardens. They provide it all – shelter, structure, diversity, nectar flowers, and fruit. There are many choices among indigenous species; nearly 150 species occur through the Caribbean, with over a 100 being endemic; 11 occur in Jamaica alone. All have similar wildlife values. They are shrubs or small trees about 5 m tall but are readily prunable into hedges. They can grow in the shade, along forest and garden edges, and in diverse hedgerows. Such mixed plantings of different coffee species provide diversity in flowering and fruiting. Flowers attract butterflies, bees, and other insects. The great importance of wild coffees for birds and other wildlife is their extensive crop of attractive, small fruit. Local birds seem to monitor plants for the moment a berry is ripe enough to be eaten. Birds disperse seeds widely, so wild coffees will volunteer. Wild coffees survive hurricanes well, quickly fruiting after to provide food for surviving birds. Among the more common and available are smooth wild coffee (*Psychotria ligustrifolia*) (Bahama wild coffee, Cuban wild coffee) natively from The Bahamas

Flowering wild coffee with zebra longwing butterfly.

Wild coffee fruit.

and Greater Antilles; shiny-leaf wild coffee (*Psychotria nervosa*) (balsamo, Seminole balsamo, wild coffee), occurring Caribbean-wide; hairy wild coffee (*Palicourea/Psychotria pubescens*), also occurring Caribbean-wide; Cuban wild coffee (*Psychotria cathetoneura*), endemic to Cuba; and velvet leaf wild coffee (*Psychotria sulzneri/tenuifolia*) (dull-leaf wild coffee, shortleaf wild coffee) from the Greater and Lesser Antilles. Of course, cultivated coffee (mostly *Coffea arabica*) is widely grown on suitable sites in the Caribbean. Coffee plants have wildlife value in garden and smallholding farm settings; and commercial coffee, especially shade-grown coffee, has similar wildlife values although the birds face competition for the fruit. Wild coffees are propagated by seed, transplanting, airlayering, and cuttings and are increasingly available from plant nurseries and also native plant distribution programs.

Milkweeds (Asclepiadaceae): Famously, these plants are hosts for desirable milkweed butterflies including monarchs, tigers, and queens (*Danaus*) and so are essential natural garden and butterfly garden plants. Flowers attract bees, their primary pollinators, butterflies, and wasps. The most widespread throughout the Caribbean is the tropical milkweed (*Asclepias mellodora/curassavica*) (bloodflower, bouvin, butterfly plant, herbe madame, ipeca d'Haiti, scarlet milkweed, silkweed, sunset flower). Caribbean milkweed (*Asclepias nivea*) occurs in Puerto Rico and the Virgin Islands. White twine vine (*Funastrum clausum*), from the Greater and Lesser Antilles, is a climber producing spreads of white flowers. Milkweed butterflies depend on milkweed. Sedentary monarch (*Danaus plexippus*) populations are dispersed around the Caribbean; migratory monarchs pass through northern Cuba. Jamaican monarch (*Danaus cleophile*) and tropical queen (*Danaus eresimus*) are endemics. Temperate controversies about planted milkweed interfering with monarch migration are not pertinent for the sedentary populations of the Caribbean. Milkweeds are short-lived plants, propagated by wind-dispersed seed and will volunteer. Milkweeds, especially tropical milkweed, are available in plant nurseries.

Tropical milkweed.

Bitterbush (*Rauvolfia viridis/nitida*) (bois lait femelle, bois saisiement, bwa lèt femèl, glasswood, milkbush, palo amargo, rauwolfia, smooth rauwolfia): Bitterbush is an excellent bird-plant that ranges naturally through the Caribbean and has been introduced in many other places including India and Australia. It is a 5–7-m tree occurring in wet forest, scrub, and disturbed sites. Clusters of showy, white, spring-summer flowers attract pollinators. Milky, 1-cm, 2-seeded, summer-fall, red-turning-black fruit attract birds, especially pigeons. Rauvolfia chemistry, much of it toxic, is well studied including as central nervous system depressants. In Ayurvedic medicine, rauvolfia is used for calming; in Europe, it is used for fevers, syphilis, and skin problems. Its yellow wood was used for furniture and musical instruments to the point of overharvest. It is propagated by seed and by root and stem cuttings.

Varronias (*Varronia/Cordia*): Varronias are important garden insect bushes, supporting armies of invertebrates including sulphur butterfly caterpillars. Identification of species available can be difficult as some have entered horticulture, the previous genus name of *Cordia* is often applied, and common names can be similar; but all are excellent wildlife plants. Clusters of tiny flowers attract many butterflies, including large species as well as ants, bees, native flies, and wasps, along with their predators – flycatchers, vireos, and warblers. Tiny, sweetish-pulped, 1/2-cm fruit is consumed by small- and medium-sized fruit-seeking birds. Varronias can be part of a natural garden, butterfly garden, or planted in a container, which controls their exuberant growth. Once established they will volunteer. They are salt- and drought-tolerant and so do well in plantings. They typically are used medicinally for a variety of issues including pain, circulatory concerns, and infections. Granny bush (*Varronia bahamensis*) (Bahama manjack, cocobey), from The Lucayan Archipelago and Cuba, is a multibranched shrub growing to 2–3 m, sometimes taller. Curacao bush (*Varronia bullata*) (bloodberry, butterfly bush), widespread in the Caribbean, is a mounding vine-like shrub 2–3 m or taller when climbing over other plants. Butterfly sage (*Varronia oaxacana/globosa*) (aguardientillo, bloodberry, curacao bush, Maria-preta, papita), originally a nonnative from Mexico to Central America, is a scrubby shrub growing to be 3 m tall. Varronias are propagated from seed or cuttings. Butterfly sage and granny bush are available in plant nurseries.

Geiger tree (*Cordia sebestena*) (aloewood, anacahuita, bois d'ine, bois râpe, fleurs coquelicot, kòkèliko, scarlet/orange cordia, sebesten plum, siricote blanco, ziricote): This attractive tree is natively a plant of coastal scrub and shorelines but does very well in public gardens, as street trees, in diverse natural gardens, and as stand-alone specimens. Plants grow quickly to 3–5 m, with gray, furrowed bark and a round crown, which provides cover. The sensational orange, tubular flowers attract not only insects, especially bees including honeybees but also butterflies and hummingbirds. The fruit is edible, although meant for floating. It also hosts a spectacular iridescent beetle. It is traditionally used for treating infections. Bocote (*Cordia alliodora*) (bois soumis, capa, freijo, jennywood, manjack, Spanish elm, varia), from the Greater and Lesser Antilles, is a tree up to 20 m tall, used for lumber and in coffee plantations. Geiger trees are readily propagated from seed, by transplanting seedlings that emerge under mother plants, and are available in plant nurseries.

Rooster comb (*Heliotropium angiospermum*) (crete coq, herbe à malingres, scorpiontail, zimbaclot): This hardy, short-lived perennial is a small shrub, growing up to 1 m, naturally widely distributed in the Caribbean not only in scrub forest edge but also in disturbed

Geiger tree.

areas and roadsides. Multiple, tiny, 1-mm white flowers on curling spikes, despite their small size, attract a surprising number and variety of insects. It is used across the Caribbean for various skin issues such as diaper rash, prickly heat, and measles. Quite weedy, rooster comb can be started from seeds or transplanted from disturbed sites and, once in the garden, will volunteer.

Bourrerias (*Bourreria*): Bourreria species are found diversely throughout the Caribbean, but taxonomy is a bit confusing as species limits change with research, and they share common names. The most common and widespread, *Bourreria succulenta*, now is considered to include previously recognized parrotberry (*Bourreria venosa*), Bahama strongback (*Bourreria ovata*), and the formerly Jamaican endemic *Bourreria velutina*. Taxonomy lessons aside, bourrerias are important wildlife plants for their abundant flowers and fruit. *Bourreria succulenta*, most commonly called strongback in English (acomat cotelette, Bahama strongback, bodywood, bois cabrit batard, bois puant, cafe mawon, cherry, chink, common manjack, Madame Jeanne, mapou gris, parrotberry, poisonberry, smooth strongback, strongbark, viagra banbora), occurs throughout the Caribbean. It is a shrub to small tree, of 3–8 m, with slender trunk and irregular crown. Bourreria are among the best butterfly plants. Clusters of showy, 1-cm, white, tubular flowers attract bananaquits, bees, butterflies, hummingbirds, and wasps. Orange or red 1-cm fruit, which may cover the plant, are eaten by birds, especially pigeons. Leaves and twigs and sometimes bark and roots are used alone or in a blend as a tea for general health and treating backache, bedwetting and other urinary problems, and for virility, hence some of the common names – it is considered an essential ingredient in bush teas aimed at men. Bourrerias can be used in any garden setting, including in containers, and are best for wildlife in mixed shrub settings. Island-indigenous species are best used. Bourrerias are propagated by seed and by cuttings and are increasingly available in plant nurseries.

Strongback.

Soldier vine (*Myriopus/Tournefortia volubilis*) (basket withe, bastard ratroot, chigga nut, jigger nut, kallaba, menimene, nigua, nigua de paredon, nigua enredadera, rabois, soldierbush, twining soldierbush): Found throughout the Caribbean, soldier vine is a fast-growing, perennial, vining shrub with fuzzy stems and leaves that sprawl as far as 8 m along scrub or forest edges. Long inflorescences of small, yellow flowers attract numerous insects, especially bees, as well as an abundance of spiders. Pea-sized, black-spotted, white-turning-black fruit persist on the plant and are available to birds nearly year-round. Soldier vine is used for skin problems, for coughs, and as an aphrodisiac in bush tea. Native throughout the Caribbean, it is readily found in the wild. Propagation may be from seed, cuttings, or transplanting. It is an excellent addition to a garden as an edge plant where it will readily reseed.

Nightshades (Solanaceae): From aubergine to zombies, solanacids have for millennia provided people and wildlife with food and natural products. Many species are indigenous to the Caribbean and worthy of conservation attention; others were introduced centuries ago; some are used for folk medicine; others are common edible garden crops – aubergines, peppers, potatoes, tomatoes. Interesting Caribbean natives, too many to discuss, tend to be bushes to small trees with flowers that attract bees and small tomato- or pepper-like, bird-sized fruit. Cankerberry (*Solanum bahamense*) (Bahama nightshade, berry bush, bitterberry, concaberry, corberry) is a Caribbean-wide, bushy, 1/2-m, weedy shrub having small red fruit and used as leaf tea gargled for sore throat and cankers. Gooma bush (*Solanum americanum*) (agouma, American black nightshade, dolly tomato, glossy nightshade, guma bush, ink plant, nightshade, small-flowered nightshade, spinach, zeb ame) is a highly variable, straggling, short-lived, quite toxic, weedy roadside herb, with widespread distribution in the Neotropics. Nonetheless, its young leaves are eaten like spinach. Salve bush (*Solanum erianthum*) (brier tree, mullein nightshade, potato tree) is a tall,

Cankerberry.

open-crowned shrub having attractive white flowers with yellow centers and 2-cm, yellow-green, tomato-like fruit eaten by all fruit-seeking birds. Pigeonberry (*Solanum torvum/stramoniifolium*) (gully beans, pea eggplants, prickly nightshade, shoo-shoo bush, susumber, turkey berry, wild eggplant, and many other names) is a small, native, erect, thorny bush that grows both wild and in gardens throughout the Caribbean and has become naturalized around the world's tropics. It is used in stews, but cooking is essential. Edible peppers (*Capsicum frutescens/annuum*) also are Caribbean natives. Natively, this species (or 2 species) has a small, upright fruit called a bird pepper, a name that reveals its wildlife value. It likely was brought by Indigenous People before Spanish contact. Pepper "trees" are short-lived (1–4 years), short-statured plants (1–2 m). There are many cultivated varieties including chilies and scotch bonnets. Solanacids have provided immense fodder for bush medicine, alluded to by common names. Alkaloids α-solanine and α-chaconine have been studied extensively. Selective breeding has toned down the alkaloids in cultivated solinacids; but, these plants over the centuries have been used for deliberate poison, cosmetics, folk medicines for allergy, inflammation, pain relief, muscle relaxation, modern chemical drugs, and, according to some of its names inducing zombies. Capsaicin is very commonly used to treat muscle pain through its effect on nerves. Nearly all solinacids are wildlife-friendly and should be considered for medicinal, herb, or native gardens. They are propagated by seed, cuttings, air layering, and transplanting those occurring on roadsides.

Firecracker plant (*Russelia equisetiformis*) (Antigua heather, arete panameño, coral-blow, coral fountain, coral plant, firecracker bush, fountain plant, lágrimas de Cupid, rain coral, ruselia): Originating in Mexico, this is a popular herbaceous, garden, and landscaping plant, widely used in hummingbird and butterfly gardens and in containers, which do check its spreading tendencies. About 1 m tall, its main wildlife value is its year-round, 2–3-cm, bright-red, tubular flowers that attract butterflies and hummingbirds. It is propagated by division and is available from plant nurseries; red-flowered varieties are best for birds. It is not suitable for native restorations.

White alling (*Bontia daphnoides*) (aceituna americana, mang blanc, mangle, olijfi, olivier batard, olivier de mer, ollyush, sea olive, white olive, wild dilly, wild olive, and many more): This well-known, coastal, salt-tolerant shrub or small tree, up to 9 m tall, occurs on most islands Caribbean-wide. Dense leaves in terminal clusters provide cover for birds, lizards, and insects. Year-round, yellow and pink, tubular, 1–2-cm flowers attract nectar-seeking insects and birds. Yellow drying to brown, 2-cm fruit are likely water-dispersed but also are eaten by larger fruit-eating bats, birds, and land crabs. White alling can be used as an attractive hedge or part of a mixed hedgerow, especially near the coast. It is propagated by seed and cuttings and is sometimes available from plant nurseries.

Wild petunias (*Ruellia*): These are commonly planted, short-statured plants, native on some islands. They are larval hosts for malachite and buckeye butterflies. Depending on species, flowers attract bats, bees, butterflies, or hummingbirds and so are widely planted in gardens. Mexican petunia (*Ruellia simplex/brittoniana*) (Britton's wild petunia, Mexican bluebell) is a Greater and Lesser Antilles native widely used in horticulture. Fever root (*Ruellia tuberosa*) (bluebell, cracker plant, ipeca batard, iron root, minnie root, popping pod, sheep potato, snapdragon root, toi ting, trai-no, waterkanon, watrakanu) is native through the Caribbean and has been introduced globally; in some places, it is invasive. Green shrimp plant (*Ruellia blechum*) (Brown's blechum), also a Caribbean-wide native, is a popular ornamental owing to its stacked leaf structure. Wild petunias have seed capsules that split open, which is the origin of some of its common names. Other names refer to these species' many traditional medicinal uses in various cultures for pain, muscles, bladder issues, and cough. Despite – or maybe because of – their ubiquity, wild petunias are worthwhile natural garden plants. They are propagated by seed, cuttings, and division and are available in plant nurseries.

Yellow elder (*Tecoma stans*) (chevalier, esperanza, fleur de Saint Pierre, ginger-thomas, torchwood, yellow bells, yellow trumpetbush, yellow trumpet flower): Native to the Caribbean, yellow elder generally is a fast-growing, somewhat brittle shrub 1–3 m tall, more rarely a small tree. It blooms year-round, producing fragrant, 4–6-cm, yellow, trumpet-shaped flowers that attract hummingbirds, bees, and butterflies. It is appreciated as the official flower of The Bahamas and the U.S. Virgin Islands. Naturally common in waste areas and scrub forest edges, it also is widely planted. Weedy and dry-season-tolerant, yellow elder does well in parks, natural and butterfly gardens, and patio containers. Leaves, sometimes twigs and dried bark, are used for tea generally, as a diuretic, and, famously, in folkloric blends for diabetes and high blood pressure. It is propagated by seed, keeping seeds damp, and in the light until they quickly sprout.

Tabebuias (*Tabebuia*): Two native species occur widely through the Caribbean in natural forests and also planted roadside, streetside, and in parks and gardens. They are trees to 20 m, although generally shorter, straight-trunked with attractive gray to brown, furrowed bark, and open branches that support large numbers of insects and insectivorous birds. Drought-resistant and dry-season deciduous, they bloom just before the leaves

Yellow elder.

emerge at the start of rainy season. Nectar-rich, trumpet-shaped flowers can cover the tree. White cedar (*Tabebuia heterophylla*) (pink manjack, pink poui, pink trumpet tree, powye roble blanco, whitewood) occurs Caribbean-wide. Other native tabebuias are more range-restricted, such as five finger (*Tabebuia bahamensis*) (beef bush, chicken toe, gumwood, white-cedar), a Bahamian endemic. Tabebuias are well appreciated medicinally for a range of issues and as an ingredient in aphrodisiac teas. Tabebuia wood is termite- and rot-resistant. Central and South American relatives, known as ipe (*Handroanthus* spp.), are important saw-lumber trees. Tabebuias are propagated by seed and by transplanting overly numerous volunteer seedlings.

Fiddlewood (*Citharexylum fruticosum/spinosum*) (grenarde marron, long Tom, madame claude, madam klòd, old woman's bitters, pendula, spiny fiddlewood, trash-a-berry, white fiddlewood): Occurring throughout the Caribbean, fiddlewood can be tall trees to 15 m in forest settings, short trees in secondary conditions, or kept to shrubs in natural gardens as they tolerate trimming. Suckers, if allowed, can create a thicket providing excellent structure for cover and nesting birds. Glossy leaves support an impressive community of ants, beetles, bugs, caterpillars, and termites. Leafroller caterpillars (*Epicorsia oedipodalis*) may periodically denude plants, which soon recover. Fragrant, white flowers in impressive, but short-duration, drooping clusters attract butterflies, honeybees, and moths. Small, 1-cm, orange-turning-red-brown, fleshy, sweet, cherry-like fruit attract fruit-eating birds including mockingbirds and warblers especially yellowthroats. It is a useful plant for parks and natural gardens and also can be used in commercial settings. Several other native citharexylums occur on various islands. All have wildlife value. Fiddlewood is widely used for indigestion, bathing, mixed with other plants in tonics, and historically for stringed instruments and boatbuilding. This is one of the more useful plants not only for any natural planting but also for commercial settings, parking lots, roadside plantings, screens, and buffers. It can be grown from seeds or cuttings and is also available at garden shops. Several other native species of native citharexylum occur on various islands in the region, allowing the use of indigenous species for restoration and natural gardens. Hybrids, and naturalized species also occur, but in all cases natives should be preferred.

Golden dewdrop (*Duranta repens/erecta*) (azotacaballo, bois jambette, duranta, flor celeste, garbancill, jambette, mais bouilli, pigeonberry, tala blanco, skyflower): This Caribbean native is used as a garden plant in the region and elsewhere, becoming invasive in places where introduced. It is a thorny, somewhat viny, drought-tolerant shrub when pruned or a tree growing up to 10 m. Thorns provide protective cover. Purple, 1/2-cm flowers, which can occur year-round but mostly summer-fall, attract bees, birds, butterflies, moths, and wasps. It fruits simultaneously with flowers. Golden, 1-cm fruit is eaten by all fruit-eating birds, including pigeons (pigeonberry). Occurring naturally in sandy and disturbed sites, it is hardy in a natural or butterfly garden or as an attractive and thorny addition to a diverse hedgerow. Golden dewdrop is propagated by seed after processing, and by cuttings, and is available from plant nurseries.

Porterweeds (*Stachytarpheta*): Widespread, widely recognized, and diversely named throughout the Caribbean, porterweeds have been brought into gardens around the world for two centuries both for their flowers and for their medical uses. As a result, their taxonomy, descriptions, and likely genetics remain confusing. Both native and nonnative species occur in the Caribbean having similar wildlife values. They are herbs to small shrubs, 1–2 m tall depending on species, that natively are scrub-forest edge plants and can colonize harsh sites. They provide important low cover. Warblers, vireos, grassquits, and other small birds as well as lizards are attracted to hunt invertebrates under the plants. They flower on tall spikes year-round, with a few open at one time attracting a procession of bees and butterflies, including swallowtails, sulphurs, skippers, and zebra longwings and also hummingbirds and bananaquits to taller plants. Bullfinches, grassquits, and wintering warblers eat the small, 1/2-cm fruits. Blue porterweed (*Stachytarpheta jamaicensis*) (blue flower, blue snakeweed, bastard vervain, eye bright, Jamaica vervain, purple worry vine, rat's tail, venven latje, vervain, verveine caraibe, verveine queue de rat, vevenn ke a rat, worryvine) is a widespread, native garden favorite that also is very much at home on waste areas, salt flats, and cleared land. It volunteers hardily in gardens and yards and does well in containers. Other Caribbean natives include Bahama vervain (*Stachytarpheta fruticosa*) from The Bahamas and Greater Antilles; Cayenne snakeweed (*Stachytarpheta cayennensis*) (blue snakeweed, blue top, Cayenne porterweed, dar-blue snakewee, eye bright, false verbena, nettleleaf porterweed, piche de gato, rabo de zorro, rattail) from the Greater and Lesser Antilles; and coral porterweed (*Stachytarpheta mutabilis*) (changeable velvetberry, pink snakeweed, pink rat tail) from the Greater Antilles. Caribbean species have been introduced elsewhere in the tropics and in some places are invasive. Nettleleaf

Blue porterweed.

velvetberry (*Stachytarpheta urticifolia*) may have originated in Asia but has been spreading widely including in the Caribbean. Porterweeds are used medicinally by many cultures, mostly as foamy leaf teas (thus the name) for treating fever, colds, constipation, skin issues, and worms. Blue porterweed has proven effectiveness for pain and inflammation, validating some traditional uses. Although sometimes it is hard to identify native species from imports, using species indigenous to an island provides conservation value. They are short-lived, propagated by dried seed, cuttings, and transplanting seedling volunteers. Porterweeds are also available in plant nurseries; but sorting out native species takes some study.

Sages (*Lantana*) (lantanas, verbenas): Lantanas are essential plants in Caribbean gardens, parks, and restoration sites for their many wildlife values. Overall, there are more than 100 species of lantanas and many hybrids; their taxonomy is difficult, even contentious. Many share common names. Depending on species and growing conditions, they are ground covers to short shrubs. However, all the sages are insect plants. They are host to moth and scrub-hairstreak caterpillars. Most flower year-round attracting butterflies such as monarchs, skippers, sulphurs, swallowtails, julia, bees including honeybees, and other nectar-seeking insects. Taller plants are used by bananaquits and hummingbirds. Their small fruit, likewise available through much of the year, is eaten by many birds especially doves and pigeons and also mockingbirds, vireos, and warblers, including the endangered Kirtland's warbler in the Bahamas. Butterfly bush (*Lantana involucrata*) (big sage, button sage, cariaquito blanco, cariaquito encarnado, cerise, native sage, West Indian sage, white sage, wild lantana, wild sage), native throughout the Caribbean, is an essential plant in a tropical butterfly-friendly garden. Wild sage (*Lantana camara*) (ham-and-egg, lantana, largeleaf lantana, red sage, shrub verben, Spanish flag, West Indian lantana, West Indian sage) is also a well-known Caribbean native that is widespread and highly invasive where introduced. Other potential garden and restoration species are yellow sage (*Lantana bahamensis*) (wild sage) from The Bahamas and Greater Antilles; Bahama sagebrush (*Lantana demutata*) (tiny sage) from The Bahamas; shrub verbena (*Lantana exarata*) from the Greater Antilles; and rock sage (*Lantana reticulata*) (measle bush, sage) from the Greater and Lesser Antilles. Lantanas are used for treating many traditional medical issues such as blood pressure and especially skin conditions. These are tough, weedy plants, easily propagated by seed after soaking in water for a day and by transplanting seedlings. Birds volunteer them. Some are available in plant nurseries, preferably selecting for indigenous species.

Bushy lippia (*Lippia alba*) (brisée trois tasses, hierba dulce, juanilama, karabo, licorice verbena, lippia, Marguerite, Marguerite blanche, Marguerite des jardins, pampo-regano, poleo, pitiona, twa tas): Caribbean-wide native, cultivated for attractive flowers and aromatic foliage. It is a short, 2-m, multibranched shrub. Its white, pink, or light blue-purple flowers on spikes along branches attract bees, butterflies, and other nectar-seeking insects. Although native to the Caribbean, it is introduced elsewhere. It is one of the more widely used native plants wherever it is found, including as a food flavoring and medicinally as a sedative and antidepressant. Lippia is propagated primarily by cuttings with several leaves left on the stem and by transplanting volunteers.

Fogfruit (*Phyla nodiflora*) (common lippia, frogfruit, turkey tangle): Widespread naturally in the New World and introduced elsewhere, this low (up to 6 cm tall) spreading

Butterfly bush with long-tailed skipper.

Fogfruit.

ground cover can be an important contributor to a diverse lawn. It is a larval host for buckeye, crescent, and peacock butterflies, and its tiny purple and white flowers are nectar sources for bees, butterflies, and other insects. It is essential in untrampled diverse lawns, where it is mowable at high settings. Propagation is by division.

Tropical sage (*Salvia coccinea*) (blood sage, crimson sage, hummingbird sage, red flowering salvia, red sage, red salvia, rosy sage, scarlet sage, scarlet salvia, South American sage, Texas sage, wild salvia): This originally Caribbean native is widely distributed in horticulture. West Indian sage (*Salvia occidentalis*) is a native but with a confusing taxonomy. Originally from the Greater and Lesser Antilles, it has been introduced around the world where in places it has become a troublesome invasive. Sages are short-statured to about 1 m with a lanky, upright growth form; their attractive red flowers clustering at the tips of stems make them readily available to pollinators, which cannot seem to resist. Flowering through much of year, especially summer and autumn, it is continuously visited by bees, butterflies, and hummingbirds. It is an essential plant in butterfly or hummingbird gardens, used generally at edges or along paths. It is widely cultivated with various color varieties available; but for hummingbirds, the original red is the best. Sages are short-lived, but, once in a garden, they self-seed. Propagation is by seed and by transplanting seedlings but easily from cuttings. They are available in plant nurseries.

Bastard stopper (*Petitia domingensis/urbanii*) (bois d'ortie, bwa dòti, capá, chen kalbas, fowlberry, guayo, pepperberry, wild guava, zombia): Being a mint-relative widespread species through the Caribbean, fowlberry is a shrub or tree, 3–8 m (but prunable), having large 20-cm leaves creating a spreading but open crown that provides ample cover and structure for birds and other animals. It occurs in diverse habitats and is not particular as to soil or water conditions. In spring and summer, fragrant, 1-mm flowers attract insects, especially honeybees. The red, 1/2-cm, summer-fall fruit cluster at branch end making them readily accessible to birds. Leaf teas are used for treating digestive and respiratory issues and for fever. Fowlberry has furniture- and tool-quality wood. It is valuable in ecosystem restorations and any other natural planting. It is propagated by seed and transplanting seedlings before tap roots grow.

Inkberry (*Scaevola plumieri*) (beachberry, fanflower, gulfweed, pen & ink, waxybush): This native shrub is a common dune plant and important dune stabilizer throughout the Caribbean. Its dense leaves on multiple stems form thick, bushy clumps providing excellent shelter. Native inkberry is a hardy, salt-tolerant, sandy- and dry-soil plant that does well in restorations, parks, sandy gardens, parking lots, and commercial sites. Its distinctive, 2-cm white flower looks like a half flower. The dark purple, 2-cm berry (an old-time ink substitute) floats for coastal dispersion but it is also eaten by birds and iguanas. An aggressive, nonnative relative, Hawaiian seagrape or beach naupaka (*Scaevola taccada*) is on the do-not-plant list because it colonizes beachfronts smothering native plants. Native inkberry can be propagated by seed, cuttings, and burying branch tips in the sand until they root.

Spanish needle (*Bidens pilosa/alba*) (beggar's tick, blackjack, burr marigold, margarita silvestre, romerillo, shepherd's needle, tickseed): One of most common native asters through the Caribbean, Spanish needle is a short-lived, rather scraggly, short-statured plant, 1/3–1 m tall. It is so common that its wildlife value may go unappreciated. White and yellow, daisy-like flowers attract bees, butterflies, moths, and other insects. Seeds that do not attach to trousers are eaten by small seed-eating birds. It is a hardy, weedy plant of waste places and roadsides and essential in native and butterfly gardens, whether it volunteers or not. Leaves are edible and as a tea are used for treating blood pressure and colds. Crushed leaves are used for skin issues. It has been scientifically shown to have antibacterial, antifungal, antidiarrheal, antioxidant, and antihypertensive properties. It is propagated by seed and transplanting and may aggressively volunteer.

Spanish needle with hammock skipper.

Shrubby sea oxeye daisy.

Shrubby sea oxeye daisy (*Borrichia arborescens*) (bay candlewood, green sea oxeye daisy, tree seaside tansy, tree seaside oxeye): The Caribbean native sunflower is a coastal wetland shrub that forms dense, multi-stemmed, 2 m tall stands. Its fleshy leaves and dense stems provide cover. Yellow, 2-cm, daisy-like flowers bloom year-round, although peaking in summer, attracting butterflies. Bullfinches, seedeaters, and sparrows eat the seeds. It is salt-tolerant but shade-intolerant. It is versatile in butterfly and native gardens and in sunny locations along decorative edges or as part of a short diverse hedge. It spreads by rhizomes and can be propagated by seed, cuttings, and rhizome division. It may be available in plant nurseries or from native plant distribution programs.

Beach ragweed (*Ambrosia hispida*) (bay geranium, bay gerina, bayjareena wormwood, bay tansy, bay thyme, bay vine, coastal ragweed, soap bush, sweet bay): Native in sandy beach areas throughout the Caribbean, this crawling, vining plant with hairy, gray leaves provides low cover. Small 1/2-cm nut-like fruit are eaten by seed-eating birds. It is an important plant in folk medicine, thus all the names. It is used for treating worms (especially at the start of school year), colds, fever, and high blood pressure and as soap for skin problems. Beach ragweed tolerates dry, salt, and wind and so is hardy in plantings. It is propagated by seed and transplanting.

Matchwood (*Schefflera/Didymopanax morototoni*) (aralie grandes feuilles, arriero, bois canot, gorgorán, jeretón, morototo yagrumo macho, mountain trumpet, palo de sabre, sablito, yagrumo macho, zapaton): Widespread in the Greater and Lesser Antilles and elsewhere in the Neotropics, this native, fast-growing, pioneer tree, 8–30 m tall, colonizes disturbed sites. It has a straight trunk with ornamental whirls of leaves forming a dense, umbrella-like crown. Flowers attract bees; and the fruit is an important food resource for bats and fruit-eating birds. It is used medicinally for malaria, wounds, rheumatism, and issues of bone, joint, and skin. It is a timber tree, especially suitable for making furniture and boxes. It is valuable in early restorations and native gardens. The related umbrella tree (*Schefflera actinophylla/Heptapleurum actinophyllum*), from New Guinea and Australia, is on the do-not-plant list. Matchwood is propagated with difficulty by scarified seeds or better by transplanting seedlings.

References

1. Christenhusz, M. J. M., J. L., Reveal, A. Farjon, M. F. Gardner, R. R. Mill, and M. W. Chase. 2011. "A New Classification and Linear Sequence of Extant Gymnosperms." *Phytotaxa* 19:55–70.
2. Angiosperm Phylogeny Group. 2016. "An Update of the Angiosperm Phylogeny Group Classification for the Orders and Families of Flowering Plants: APG IV." *Botanical Journal of the Linnean Society* 181:1–20. DOI:10.111/boj.12385
3. Adams, C. D. 1972. *Flowering Plants of Jamaica*. Kingston: University of the West Indies.
4. Little, Jr., E. L., R. O. Woodbury, and F. H. Wadsworth. 1974. *Trees of Puerto Rico and the Virgin Islands*. Agricultural Handbook 449. Washington, DC: U.S. Department of Agriculture, Forest Service.
5. Honychurch, P. N. 1980. *Caribbean Wild Plants and Their Uses*. Oxford: Macmillan Education.
6. Scurlock, J. P. 1987. *Native Trees and Shrubs of the Florida Keys*. Bethel Park: Laurel Press.
7. Nellis, D. W. 1994. *Seashore Plants of South Florida and the Caribbean*. Sarasota: Pineapple Press.
8. Light, M. H. S. 1995. *Growing Orchids in the Caribbean*. Oxford: Macmillan Education.
9. Timyan, J. 1996. *Bwa Yo: Important Trees of Haiti*. Washington: South-East Consortium for International Development.
10. Carrington, S. 1997. *Wild Plants of the Eastern Caribbean*. Oxford: Macmillan Caribbean.
11. Osorio, R. A. 2001. *Gardener's Guide to Florida's Native Plants*. Gainesville: University Press of Florida.
12. Patterson, J. 2002. *Native Trees of the Bahamas*. Nassau: Bahamas National Trust.
13. Cutts, W. 2004. *Trees of The Bahamas and Florida*. Oxford: Macmillan Education.
14. Gann G. D., and J. C. Trejo-Torres. 2005–2023. *Plantas de la Isla de Puerto Rico*. Delray Beach: Institute for Regional Conservation. www.regionalconservation.org/ircs/database/site/IntroPR.asp
15. Gann G. D., C. J. Abbott, C. G. Stocking, K. N. Hines, and collaborators. 2005–2023. *Natives for Your Neighborhood*. Delray Beach: Institute for Regional Conservation. regionalconservation.org/beta/nfyn/default.asp
16. Kass, L. B., A. J. Kowalski, and R. E. Hunt. 2009. *An Illustrated Guide to the Common Plants of San Salvador Island, Bahamas*, 3rd edition. San Salvador Island: L. H. Bailey Hortorium and Grace Research Centre.
17. Acevedo-Rodríguez, P., and M. T. Strong. 2012. "Introduction." In *Flora of the West Indies*. National Museum of Natural History. www.naturalhistory2.si.edu/botany/WestIndies
18. Kushlan, J. A., and K. Hines. 2014. *Attracting Birds to South Florida Gardens*. Gainesville: University Press of Florida.
19. Lindsay, K. C. 2014. "The Ferns of Antigua and Barbuda: A Case of Resurgence and Resilience." *Pteridologist* 6:14–18.
20. Currie, D., J. M. Wunderle, E. Freid, D. N. Ewertm, and D. J. Lodge. 2019. *The Natural History of the Bahamas: A Field Guide*. Ithaca: Comstock Publishing Associates.

21. Field Museum. 2023. "Fern Genera of Cuba." www.fieldguides.fieldmuseum.org/guides/guide/198
22. Leon Levy Preserve. 2023. "Leon Levy Preserve Leon Levy Native Plant Preserve." www.levypreserve.org/flora-by-common-name/
23. National Museum of Natural History. 2023. *Plants of the West Indies.* Washington, DC: Smithsonian Institution. www.naturalhistory2.si.edu/botany/WestIndies/
24. University of West Indies. 2023. "Plants of the Eastern Caribbean." www.ecflora.cavehill.uwi.edu
25. BirdsCaribbean. date unknown. "Caribbean Heritage Plants." www.birdscaribbean.org/wp-content/uploads/2014/06/Heritage-Plants-BirdsCaribbean.pdf
26. Sánchez, C. 2021. "Inventario de los Licófitos y Helechos de Cuba." *Revista del Jardín Botánico Nacional* 42:1–53.

6 Birds, butterflies, and other animals

A goal of recovering nature whether in large reserves, urban parks, or small private gardens is fundamentally accomplished through creating and maintaining large and small communities of native plants that in turn provide habitat for indigenous bees, birds, butterflies, frogs, lizards, and many other animals, preferably including as wide a spectrum of native animals as natural and feasible for the location. The larger and more connected the restoration, the more species and numbers of animals it can support. Contrasted with large reserves, recovering nature on smaller scales, such as public parks and private gardens, inevitably results in less complete animal communities. But they still create viable and valuable habitat. Every hectare, or part of a hectare, counts. Even the smallest of private gardens or containers can support native animals. As time passes, the cumulative number of species that have occurred in such patches increases. For the most part "if you build it right, they will come," if they can get there. Getting there is more likely if all such restorations, big and small, have connectivity.

Not all indigenous animal species can be expected to frequent any particular recovered area. Each species is attracted, or not attracted, to specific characteristics an area might offer in food, nesting sites, water, cover, shelter, protection, or some combination. Restorations near the coast may attract animals not found inland; those in the mountains may attract species not present on the coast. Some animals are generalists and readily visit or reside in parks, towns, and native gardens. Some thrive in the human-dominated environment. Others may use a different sort of habitat than a park or private garden provides; xeric habitat species would not be expected in a rainforest. Most of the common garden bird and butterfly species are generalists from the coastal lowland scrub and forest habitat and can be attracted to small-scale natural gardens. Most montane species are more specialized in requiring large tracts of natural forest habitat. They can benefit from large-scale conservation and ecosystem restoration. Some have specialized needs; woodpeckers need cavity trees; grassquits need seeds; hummingbirds need flowers. As the indigenous faunas of different Caribbean islands and parts of islands differ, so do prospective species that might be accommodated in a restoration.

Following is a discussion of target animals that might be attracted to appropriately configured natural-area restorations. A large-scale restoration would aim to attract its complete complex of indigenous animals, some of which may occur only in large tracts of natural habitat. Smaller areas are more limited in what species might be attracted.

DOI: 10.1201/9781003396321-7

Given that large-scale restorations are meant to attract their specialized array of species indigenous to the island, the following discussion tends to focus on animals and groups of species that might be attracted to smaller-scale, more generalized areas such as communities, botanic gardens, public parks, and private gardens. As is the case in Chapter 5 for plants, wide-ranging, tolerant, flexible species tend to be highlighted over limited-range, habitat-restricted, or endemic species. The discussion starts noting some of the common nonnative animals found in the Caribbean, which may not be welcome in recovered habitats.

There are a number of sources for common and scientific names, distribution, and biology of the Caribbean animals.[1–26] As is the case for plants, common names differ among islands and languages, but non-exhaustive examples are provided in English, Spanish, French, Dutch, and Haitian Creole where appropriate. There are more. For birds, where possible, the principal source for the region, *Birds of the West Indies*, is followed, which includes a thorough catalogue of local Caribbean names that can be consulted.[1] Order of presentation follows the IOC World Bird List.[2] Other groups follow various accepted sources, cited earlier. The index can be used to find species accounts of interest.

6.1 Some nonnative animals

As is the case with plants, nonnative animal species can seriously affect natural and recovered areas, impacting both indigenous plants and animals.[27,28] Some can be devastating. The small Indian mongoose for many decades has been an existential threat to ground-dwelling native birds and reptiles on the many islands onto which it has been introduced. Cane toads were meant to reduce pests in sugarcane but eat native insects, other frogs, and lizards. There is hardly a more Caribbean-wide threat than a feral or out-of-doors domestic cat. But, not all nonnative species in all situations threaten Caribbean nature recovery. The Brahminy blind snake is a good example; no one realizes it is in the flower pot.

Eurasian collared-dove (*Streptopelia decaocto*) (tórtola, tourterelle turque): Old World pigeon that is an active invader in The Bahamas, Cuba, and parts of Lesser Antilles.[29] It can be abundant in developed areas. As a seedeater, it is adept at using and monopolizing seed feeders, displacing native species.

Rock pigeon (*Columba livia*) (rock dove, paloma bravia, pijon): The introduced, locally common urban street pigeon frequents public parks and private gardens, especially if fed, and nests in or on buildings. Generally, rock pigeons will outcompete native doves and, where overabundant in parks, may create health sanitation issues. Racing pigeons and escaped pets add to feral populations. Given widespread familiarity, many people like rock pigeons, which may help spark an interest in wild birds.

Parrots and parakeets (Psittacidae): Although there are parrots and parakeets native to the Caribbean, some are not, and being hardy survivors, these are usually more obvious in town than native species. Any feral or escaped parrot might show up in parks and gardens, especially those with suitable fruit trees or feeders. Brown-throated parakeet (*Eupsittula pertinax*), native to the Leeward Antilles, has been introduced to the Virgin Islands and

Rock pigeon on beach.

other places. Other successful, nonnative introduced species include the white-winged parakeet (*Brotogeris versicolurus*) (Puerto Rico); green-rumped parrotlet (*Forpus passerinus*) (Barbados, Jamaica); and olive-throated parakeet (*Eupsittula nana*) (Dominican Republic) (olive-throated conure in aviculture). The monk parakeet (*Myiopsitta monachus*) (Puerto Rico, Cayman Islands) is an urban and suburban bird that builds large communal nests that can knock out power grids. These parrots and parakeets will come to feeders, fruit trees, and agricultural crops. Native parrot and parakeet species are of conservation concern, but inviting nonnative species to a garden is not helpful in supporting indigenous birds.

Starling and mynas: European starling (*Sturnus vulgaris*), hill myna (*Gracula religiosa*), and white-vented myna (*Acridotheres javanicus*) are nonnative species introduced in various places around the Caribbean. Starlings are common in parts of Jamaica and The Bahamas. They are birds of fields, grassy areas in parks, gardens, and commercial settings, especially take-away restaurants. Mynas are aggressive birds that outcompete native species for cavity-nesting sites and feeding sites.[29,30]

Finches (Estrildidae plus): An ever-changing array of nonnative birds commonly called finches occur on various islands, having escaped aviculture. These include saffron finch (*Sicalis flaveola*) (a tanager), bronze mannikin (*Spermestes cucullata*), Indian silverbill (*Eudice malabarica*), scaly-breasted munia (*Lonchura punctulata*) (damero), and tricolored munia (*Lonchura malacca*) (monja tricolor). All are seedeaters attracted to grassy areas, agricultural areas especially sugarcane, parks, gardens, and seed feeders. European house sparrow (*Paser domesticus*) is widespread and locally abundant through the Caribbean.

Saffron finches on Bonaire.

It is an urban, garden, and yard bird that comes to seed feeders and outcompetes other species for cavity nest sites.

Green iguana (*Iguana iguana*): Native to some islands, conservation of genetically distinctive endemic green iguana populations is a matter of grave conservation concern. Green iguanas from other sources have been introduced widely through the Caribbean including to Antigua, The Bahamas, Barbuda, British Virgin Islands, Cayman Islands, Guadeloupe, Grand Cayman, Martinique, Puerto Rico, Saint Martin, and the U.S. Virgin Islands.[31,32] They colonized Anguilla from Saint Martin. These green iguanas hybridize with the endangered Lesser Antillean Iguana (*Iguana delicatissima*) and endemic green iguana populations, including on Grenada, Montserrat, Saba, Saint Lucia, and Saint Vincent. A green iguana is seldom a gardener's friend; but, perhaps even more importantly, where nonnative they are a threat to an island's genetic biodiversity.

Cane toad (*Rhinella/Bufo marina*) (crapaud géant, Cuban toad, Dominican toad, giant toad, marine toad, sapo de caña, sapo neotropical gigante, South American cane toad, spring chicken): Native to Central and South America but introduced throughout much of the Caribbean including Antigua, Aruba, Barbados, Cuba, Dominica, Grenada, Guadeloupe, Grand Cayman, Hispaniola, Jamaica, Martinique, Montserrat, Nevis, Saint Kitts, Saint Lucia, Saint Vincent, Tortola, and more widely. [33,34] The species was introduced for predator control in sugar plantations throughout the tropics, and indeed it is a significant predator of insects – although not necessarily only of cane pests – but also of other frogs, ground lizards, and small birds, as well as eating dog food, table scraps, and other scavengings. Its poison glands pose a threat to dogs.

Vervet monkey (*Chlorocebus aethiops*) (African green monkey): One of the oldest invasive introductions, going back over 300 years, now on Barbados, Nevis, Saint Christopher, and Sint Maarten. These monkeys are known to adversely affect native birds and reptiles, critically in Saint Christopher's Central Forest Reserve National Park. They also cause agricultural and horticultural damage and can be dangerous. Many control measures have been tried.[35] On the other hand, they are a significant tourist attraction on Saint Kitts and

Cane toad eating a native ground lizard in Grenada.

Barbados. Other primate introductions have also occurred in the Caribbean; none ended up well.

Small Indian mongoose (*Herpestes auropunctatus/javanicus*) (Asian mongoose, Javan mongoose): Introduced for rat and snake control in sugarcane fields as early as the late 1800s and then further introduced on other islands, mongooses now range widely in agricultural areas, towns, public parks, and private gardens. They have had a hugely negative influence on Caribbean biodiversity, especially on ground-dwelling and ground-nesting birds as well as frogs, lizards, and snakes.[36] Eradication has proven challenging and likely is impossible on large islands. Trapping has been effective on small islands and to some extent in individual parks and gardens.

Raccoon (*Procyon lotor*): Occurring scattered throughout the Caribbean and sometimes locally considered native, all raccoons are descendants of animals introduced from North America. Like their ancestors, they are very comfortable and disruptive in park and garden settings.

Pigs (*Sus scrofa*): Invasive European wild pigs have been in the Caribbean for over 500 years, having been brought there by Europeans, starting on Columbus' second voyage.[37–40] The adverse agricultural impacts of escaped hogs were soon realized, so control measures also began about 500 years ago, but their environmentally destructive ways continue, including in reserves and other natural areas. They destroy vegetation and soils and are a threat to ground-dwelling animals such as ground-nesting Bahamian parrots. Feral populations persist throughout much of the Caribbean, including in The Bahamas, Barbuda, Bonaire, Cuba, Curaçao, Dominica, Hispaniola, Jamaica, Montserrat, Puerto Rico, Saint Croix, Sint Eustatius, and the U.S. Virgin Islands. They are appreciated by

Small Indian mongoose in Jamaica.

hunters who kill them for food and sport and by tourists on Bahamian beaches. Pork, not necessarily wild pork, is a cultural staple for many island cultures, and pork production is needed to support island food security. But there is a continuing need for the mitigation of environmentally destructive impacts of wild hogs and unconfined-ranging pigs in natural, protected, and restored habitat.

Goats (*Capra hircus*): Goats, as well as burros, sheep, horses, and cattle, can be destructive to vegetation and the native animals it supports. Their grazing and browsing on the limited land base of Caribbean islands can strip natural habitat and also any other place they are released uncontrolled. Goats are illustrative. The potential environmental impact of goats depends on time and place. It is not a minor matter; there are an estimated 3.5 million goats in 15 CARICOM countries.[41] Their browsing alters wooded habitats, and their trampling dislodges soil and kills seedlings and other small plants. Goats denuded Redonda (Antigua and Barbuda) threatening the endemic Redonda ameiva (*Pholidoscelis atratus*), Redonda anole (*Anolis nubilis*), and nesting seabirds; but the vegetation made a remarkable recovery after they were removed.[42] Other eradication programs have taken place on Saba's Mount Scenery National Park and Bonaire's Washington Slagbaai Park. Controlled foraging has been recognized as enhancing wintering habitat for Kirtland's warbler in The Bahamas.[43] Mutton is a staple of many Caribbean cuisines; goat stew is the national dish of Montserrat and nearly so in Saint Christopher and Jamaica. Goat raising is encouraged by governments for food security.[44] However valuable for other purposes, the destruction that goats can bring to natural habitat is significant and goat husbandry requires management.

Rats (*Rattus*): Caribbean islands' indigenous rice rats (Oryzomyini) are extinct, the last just over a century ago. Old-World brown rat (*Rattus norvegicus*) (Norwegian rat) and black rat (*Rattus rattus*) (roof rat, ship rat) are now ubiquitous. Apart from health risks, nonnative rats' effects on wildlife have been profound, killing lizards, snakes, ground-nesting birds and their eggs, invertebrates, and plants.[45] Perhaps surprisingly, house mice (*Mus musculus*) can have similar impacts. Programs have successfully eradicated rats and mice on small wildlife-critical islands such as Dog Island (Anguilla) and Redonda and Green Islands (Antigua and Barbuda).[46–48] Plants and native wildlife responded quickly. Rats on populated islands cause substantial and recurring damage to vegetation and wildlife but are more difficult to deal with than on small uninhabited cays. They also impact parks and gardens by eating fruits, vegetables, and other plants. Control in such small holdings is difficult, but their damage can be mitigated to some extent, as discussed in Chapter 4.

Cats (*Felis catus*): Cats kill birds, lizards, frogs, and about any other small animal they can get their teeth into. Cats kept indoors are welcome pets; cats allowed out-of-doors kill. In the United States alone, cats kill approximately 2.4 billion birds every year, the largest source of direct, human-caused mortality.[49–51] Worldwide, cats have caused the extinction of over 60 bird species, mostly on islands just like those in the Caribbean. Sources of roaming cats are pets let out of doors and feral animals, which often form destructive colonies. Feral cats are also a public health concern. For small sites, high fences, thorny hedgerows, and protective dogs can help discourage outdoor cats from invading natural gardens. Residents keeping their cats indoors will do more.

Dogs (*Canis familiaris*): Feral or stray dogs are a common sight throughout the Caribbean, earning names such as potcake, satos, street dogs, or hooligans. Control is an issue

Stray dogs eating endangered leatherback sea turtle eggs they dug up from a nest.

in most places, importantly in Aruba, The Bahamas, Cuba, Dominican Republic, Haiti, Jamaica, and Puerto Rico. Free-ranging dogs' hunting, scavenging, chasing, pack behavior, and digging can cause difficulties in protected areas and natural gardens, and they may kill nesting birds and their chicks. Issues with feral and unrestrained pet dogs include disturbance and predation of birds and other wildlife, human health and safety, humane treatment, economic costs, and impact on tourism. Dogs need to be restrained.

Giant African snail (*Lissachatina fulica*): An East African native introduced to Cuba, Hispaniola, and Lesser Antilles, this destructive snail regularly shows up elsewhere. This snail can grow to 20 cm and weigh as much as a kilogram, although most are much smaller. They can devastate native plants, agricultural crops, and edible and decorative gardens.[52,53] Eradication programs, such as on Sint Eustatius and Barbados, aim to curtail the impact. Some native birds, such as limpkins (*Aramus guarauna*), have learned to eat them.

6.2 Birds

Yellow-crowned night-heron (*Nyctanassa violacea*) (bihoreau violacé, coq de nuit, crab-catcher, crabier, crabier gris, gaulin, guanabá real, quok, rey congo): This conspicuous heron is widespread through the Caribbean and is not at all shy about coming into gardens, parks, and other landscaping, particularly near the coast. Although it specializes in catching and eating crabs, it also hunts lizards and insects on lawns and in parks and gardens.

Cattle egret (*Bubulcus ibis*) (cowbird, garcita beuyera, garza canadera, gaulin, héron garde-boeufs, krabye gad-bèf, oiseau détiqueur): Cattle egrets occur throughout the Caribbean and, as their common names suggest, characteristically use pastures and fields, often although not always associating with grazing stock. They also follow tractors, lawn mowers, canecutters, and other activities that disturb insects, especially their primary food, grasshoppers. In more developed areas, they forage along roadsides and on lawns in parks and gardens.

Nightjars (Caprimulgidae): Nocturnal, nightjars and nighthawks are heard and seen mostly in the evening and at night. They are highly cryptic and keep well hidden by day in forests but also in sufficiently dense vegetation in parks and private gardens. The Greater Antillean nightjar (*Caprimulgus cubanensis*), sometimes divided into two species, may be found broadly in forest habitat including well-treed gardens in Cuba and Hispaniola. Other nightjars occur elsewhere in the region. Nighthawks (*Chordeiles*) are a common sight and sound flying over pastures, lawns, fields, parks, and residential areas, often quartering past evening street and park lights. They use trees, fences, and protected ground-sites for roosting including in parks and wooded gardens. Antillean nighthawks (*Chordeiles gundlachii*) (añapero querequeté, mosquito hawk, petonvwa, piramidig, pir-ra-ma-dick, querebebé, querequeté, rickery dick) are summer residents, nesting on the ground in protected sites including in gardens and on flat roofs. Common nighthawks (*Chordeiles minor*) (añapero yanqui, petonvwa etranje) pass through during migration. Chuck-will's-widow (*Antrostomus carolinensis*) (chouette petonvwa, don juan, guabairo americano, guabairo major) is a secretive, winter resident, mostly in Hispaniola and The Bahamas. It roosts during day in thick shrubbery including in gardens and parks.

Antillean nighthawk.

Swifts (Apodidae): Swifts forage in the air over fields, public parks, agricultural areas, and towns, sometimes accompanying swallows. Seven swift species occur in the Caribbean. The Antillean palm-swift (*Tachornis phoenicobia*) (jòljòl, vencejillo antillano, vencejito de palma) occurs in the Greater Antilles. It is common in and over villages and cities and roosting and nesting colonially in dense dead palm thatch, such as in *Coccothrinax* palms.

Hummingbirds (Trochilidae): The Caribbean is a land of hummingbirds. Of 18 Caribbean species, 16 species and 5 genera are endemic to the region. Most are single island endemics. Several occur in just 2 or 3 countries/islands or range from Puerto Rico through the Lesser Antilles. Some are found high on mountains, some along the shore; some occur on the wet side, some on the dry; most are common or fairly common, only one is of conservation concern (the bee hummingbird in Cuba); some specialize in flowers having a specific morphology, others are generalists. Positive conservation status for most species is due in large part to having available habitats with appropriate flowers, many of which are noted in Chapter 5. Hummingbird flowers are also found in dedicated hummingbird and butterfly gardens; cacao, coffee, nutmeg, and banana plantations; and artificial feeding stations. Hummingbirds consume not only nectar but also pollen and small invertebrates such as spiders. As discussed, sugar-water feeders do not provide a complete diet but do bring them out into the open to be observed and can be valuable when flower nectar is unavailable, such as after storms or in the early morning. Among hummingbirds that tend to come to parks and gardens are Cuban emerald (*Riccordia ricordii*) (zanzún); ruby-throated hummingbird (*Archilochus colubris*), bee hummingbird (*Mellisuga helenae*); Antillean crested hummingbird (*Orthorhyncus cristatus*); mangos (*Anthracothorax*); caribs (*Eulampis*); red-billed streamertail (*Trochilus polytmus*) (doctorbird, scissortail); and Bahama woodstar (*Nesophlox evelynae*) (Bahama hummingbird, hummer). Each location

Red-billed streamertail.

has its short list of indigenous species; so, a first step in determining how to attract and provide for hummingbirds is to determine which species occur on an island, in what habitats, what time of year, and what plants they feed on. For any natural plantings, from large restorations to natural gardens, site-appropriate and hummingbird-species appropriate nectar plants are main enhancements. After hurricanes, sugar-water feeding is helpful, perhaps essential, to reduce mortality.

Cuckoos (Cuculidae): Cuckoos are interesting birds that tend to move along tree branches or under bushes in search of lizards and large insects, usually keeping out of sight. Nine species occur in the Caribbean; all benefit from restorations; and a few may be expected to show up in parks and gardens. The mangrove cuckoo (*Coccyzus minor*) (cuclillo manglero, ti tako) has the largest regional breeding range and the greatest diversity in habitat use, including scrub and similarly densely planted gardens nearby. The yellow-billed Cuckoo (*Coccyzus americanus*) (coulicou a bec jaune, May bird, pájero bobo, primavera, rain bird, ti tako vant blan) is solely a summer, rainy-season resident, although North American nesting birds also migrate through. They will come to thick gardens and scrub forest. Chestnut-bellied cuckoo (*Coccyzus pluvialis*) (May bird, rain bird, old man bird) is a Jamaican endemic. The great lizard-cuckoo (*Coccyzus merlini*) (arriero, guacaica, pájaro bobo) of Cuba, The Bahamas, and Puerto Rico can be common and quite tame, coming into thickly planted and natural gardens where it creeps through shrubs, runs on ground after prey, or perches high in a tree calling. Other Greater Antilles' lizard-cuckoos are the Puerto Rican lizard-cuckoo (*Coccyzus vieilloti*), Hispaniolan lizard-cuckoo (*Coccyzus longirostris*), Jamaican lizard-cuckoo (*Coccyzus vetula*), and the endangered bay-breasted cuckoo (*Coccyzus rufigularis*), endemic to Hispaniola.

Chestnut-bellied cuckoo in the Blue Mountains of Jamaica.

Hispaniolan lizard-cuckoo at the Isabel de Torres Botanical Garden in the Dominican Republic.

Smooth-billed ani (*Crotophaga ani*) (ani à bec, bilbitin, black arnold, black parrot, boustabak, bout de tabac, crow, garrapatero, grote ani, judío, juif, merle, merle corbeua, old arnold, old witch, tickbird): These species are widespread through the Caribbean, although patchily and inconsistently. Where they occur, anis are birds of large gardens, orchards, rural settings, and public parks especially near agricultural areas. They prefer areas with both open space and cover and will frequently come to gardens nearby, especially those with hedgerows and thick plantings. Their Caribbean distribution includes The Bahamas, Greater Antilles, Cayman Islands, and several of the Lesser Antilles. Anis are highly social, roosting, nesting, feeding, and traveling in small groups. They feed on fruits, seeds, large insects, termites, and lizards, usually on or near to the ground. They are quite noticeable when in and around a garden or neighborhood although their numbers in any one location may fluctuate markedly.

White-crowned pigeon (*Patagioenas leucocephala*) (bald pate, blue pigeon, casquito blanco, paloma cabeciblanca, paloma coronita, pigeon à couronne blanche, ramier têt-blan, torcaza cabeciblanca, white-head, witkruinduif): One of the more cherished birds of the Caribbean, in part because they are hunted and eaten. As a result, they have a fly-away wariness where hunted but they are tamer where not hunted in settlements, towns, and natural gardens. Common in summer in the northern Caribbean and Greater Antilles, they nest in colonies that may number hundreds of birds or more from which they fly long distances to feed in fruit-bearing trees, coming to forests, scrub, parks, and gardens. White-crowned pigeons are known to consume over 50 species of fruit. After breeding, they disperse widely. They are hunted as a game bird in fall through spring; in The Bahamas the daily bag limit, an eye-opening 50 birds per day.

Scaly-naped pigeon in the Grenadines.

Scaly-naped pigeon (*Patagioenas squamosa*) (blue pigeon, grote blauwe duif, mountain pigeon, paloma rubia, paloma turca, pigeon à cou rouge, ramier, ramier cou rouge, ranmye kou wouj, red head, torcaza cuellimorada, torcaza morada): Scaly-naped pigeon are widespread through much of the Caribbean other than Jamaica, although in varying abundances. They are wary where hunted, and hunting mortality is a principal conservation issue. They are primarily mountain forest birds but also occur in lowland forests, farms, public gardens, natural gardens, especially on smaller islands. They eat seeds and small fruits such as bitterbush, native figs, palms, pigeon plum, and seagrape. Populations have declined in some areas but increased in others. In the Leeward Antilles, the scaly-naped pigeon's range overlaps that of the related bare-eyed pigeon (*Patagioenas corensis*), a primarily South American species that can occur in planted areas there.

Common ground dove (*Columbina passerina*) (colombe à queue noire, duppybird, grouonia, musfuifji, rolita, tobacco dove, tojosa, zòlotan ortolan): This small ground-dwelling pigeon is common through the Caribbean. It is genetically diverse with 10 subspecies. They are tame birds attracted to edges of gardens, fields, forests, public parks, roadsides, and scrub. Ground doves eat a diversity of insects, seeds, and fruits including bloodberry, granny bush, sages, and porterweed.

Caribbean dove (*Leptotila jamaicensis*) (paloma Caribeña, paloma montaraz, red-necked dove, San Andrés dove, tórtola Caribeña): The Caribbean dove occurs natively on Jamaica, Cayman, and San Andres (each with a subspecies) and has been introduced to Nassau. It is a bird of dry forest that comes to gardens and orchards to forage on the ground or in the shrubs for seeds, small fruits, and invertebrates. It may feed under seed feeders.

Mourning dove (*Zenaida macroura*) (fifi, Florida dove, long-tail, paloma, paloma rabiche, pea dove, pintail, tórtola, tórtola rabilarga, tourterelle, toutrèl ke fine): The mourning dove is a well-known bird in The Bahamas and Greater Antilles, using a wide range of habitats, native and artificial. It is widely hunted and so seasonally skittish. When not being shot at, it frequents gardens, settlements, agricultural areas; pairs can become quite tame in their residential garden. Mourning doves feed on the ground on seeds including under seed feeders and also broadly on invertebrates and on small native fruits such as bloodberry, guineaweed, sages, porterweed, and composite seeds. They frequent seed-feeding stations, foraging on the ground or at tray feeders.

White-winged dove (*Zenaida asiatica*) (aliblanca, barbarin, lapwing, paloma aliblanca, toutrèl zèl blanch, tórtola aliblanca, white-wing): Occurring commonly in The Bahamas and Greater Antilles, white-winged dove is expanding its Caribbean range. It is a garden bird, readily visiting public parks and gardens, nesting in garden shrubs, and using hanging and tray seed feeders. It eats fruits such as bitterbush, blolly, bourreria, native fig, gumbo limbo, inkberry, mastic, poisonwood, satinleaf, and seagrape.

Zenaida doves (*Zenaida*): Zenaida dove (*Zenaida aurita*) (grosse tourterelle, mountain dove, pea dove, rolón, seaside dove, tórtola cardosantera, toutèl wouj, tourterelle, tourterelle à queue carrée, toutrelle, wood dove, zenaid treurduiff) and eared dove (*Zenaida auriculata*) (tourterelle ortolan, Trinidad ground dove) are 2 quite common, ground-feeding pigeons often in flocks that use agricultural fields, towns, public parks, resorts, and private gardens. These species usually feed on the ground on seeds and small fruits and in shrubs including blolly, bourreria, native figs, gumbo limbo, inkberry, pigeon plum, satinleaf, and seagrape. Zenaida doves are common and widespread in the northern Caribbean south through the northern Lesser Antilles, whereas the primarily South American eared dove is more abundant on southern islands.

Broad-winged Hawk (*Buteo platypterus*) (chicken-eater, chicken hawk, gavilán bobo, gree-gree, guaraguao, guaraguao de bosque, malfini, malfini gwo zèl, manger-poulet, petite buse): Occurring in Cuba and the Lesser Antilles, this forest raptor benefits from forest management and restorations. It also uses parks, well-treed natural gardens, farms, and towns, capturing a wide variety of invertebrates, reptiles, frogs, and small birds.

Red-tailed Hawk (*Buteo jamaicensis*) (chicken hawk, fowl hawk, gavilán de monte, gros malfini, guaraguao, guaraguao colirrojo, malfini ke wouj, roodstaart-buizerd): This large hawk is a common bird of the northern Caribbean and Greater Antilles in sparse forest habitats, open countrysides, farms, public parks, and well-treed gardens. It nests high in trees and palms and feeds on a variety of prey including lizards, snakes, birds, and rats.

Owls (Strigiformes): The Caribbean does not lack for owls with 9 species, 5 of which are endemic. There once were large flightless owls in the Caribbean and owls figure prominently in Caribbean folklore. On some islands, they are associated with bad luck or death, transformable into witches, and killed for it. Barn owls (*Tyto alba*) (chat-haut, fresaia, frize, jumbia bird, jumpy bird, kritch owl, lechiza blanca patoo, lechuza, lechuza común, screech owl, scritch owl, white owl) and related ashy-faced owls (*Tyto glaucops*) (frize, lechuzna común) of Hispaniola are among the best known and perhaps most under-appreciated. There are as many as 6 subspecies of barn owl in the Caribbean. They use a wide range of habitats, including forests, gardens, and parks and similarly a range of nesting sites including dead trees, under rocks, caves, cliffs, broken-off palms, buildings, sheds, water towers, and specially constructed boxes. Burrowing owl (*Athene cunicularia*) (koukou, mochuelo de madriguera), occurring in The Bahamas, Cuba, and Hispaniola, nests in ground-tunnels on farms, pastures, parks, airports, golf courses, and sandy open spots in gardens. Nesting sites can be enhanced by pre-starting burrows, erecting short perches for the owls to stand on, and installing protective fencing. There are several other species of Caribbean owls: stygian owl (*Asio stygius*); short-eared owl (*Asio flammeus*); Cuban pygmy-owl (*Glaucidium siju*) (sijú platanero) and bare-legged owl (*Margarobyas lawrencii*), both Cuban endemics; Puerto Rican owl (*Gymnasio/Megascops nudipes*) (múcaro común, múcaro de Puerto Rico); and Jamaican owl (*Asio/Pseudoscops grammicus*) (potoo). These can be common in their range, except the stygian owl that is rarer and declining. Any of these owls may spend their day in trees or other thick plantings, including in rural areas and gardens. They nest in dead trees and palms in parks and well-treed gardens and may use nest boxes.

Cuban pygmy-owl.

Trogons (*Priotelus*). Two species of trogons occur in the Caribbean, the Cuban trogon (*Priotelus temnurus*) (tocororo) and Hispaniolan trogon (*Priotelus roseigaster*) (dame Angalise, kalson wouj, loro, papgayo, piragua). The Cuban trogon is the national bird, sporting the colors of the national flag and occurs throughout the island in wooded habitats. Brilliantly colored, the two Caribbean trogons occur widely in reserves, parks, and gardens where they are attracted to a wide variety of flowers, buds, insects and fruit. They call frequently but remain rather inactive, and so are observable.

Cuban trogon.

Todies (Todidae): Todies are a family of birds endemic to the Greater Antilles. The five species, including the Cuban tody (*Todus multicolor*) and Jamaican tody (*Todus todus*), have exceptionally striking plumage and a highly vocal, unafraid manner. They are habitat generalists, primarily understory birds that perch often and make small hops and flights after an array of invertebrates and also small fruit. Where sufficiently abundant and with sufficiently dense habitat, they will come to plantations and to gardens where they are highly ornamental. As they are territorial, only a pair would be expected in most garden settings.

Woodpeckers (Picidae): Twelve species of woodpeckers occur in the Caribbean, some of which readily come to well-treed parks and gardens. These include: wintering northern flickers (*Colaptes auratus*) (carpintero escapulario) and yellow-bellied sapsuckers (*Sphyrapicus varius*) (carpintero de paso); West Indian woodpecker (Melanerpes supercilliaris) (carpintero jabado, pecker-wood, redhead, red-headed woodpecker); Jamaican woodpecker (*Melanerpes radiolatus*); communally-nesting Hispaniolan woodpecker (*Melanerpes striatus*) (carpintero de La Española, sèpantye); hairy woodpecker (*Picoides villosus*) in The Bahamas; and endemic Cuban green woodpecker (*Xiphidiopicus percussus*) (carpintero verde). The Antillean piculet (*Nesoctites micromegas*) represents a genus and subfamily endemic to Hispaniola. Woodpeckers consume both invertebrates and fruit excavated from trees, palms, shrubs, bromeliads, or from the ground. Some favored insect or fruit trees include avocado, black ironwood, black olive, buttonwood, Caribbean pine, native figs, mango, palms, pigeon plum, red mangroves, seagrape, and wild tamarind. Sapsucker trees are identifiable by rings of drilled holes. The Cuban green woodpecker eats bird eggs and nectar and is a pollinator of the Geiger tree. Woodpeckers also may come to lard, peanut butter, peanuts and other nuts, fruit, suet, or sunflower seeds. They nest in cavities, especially in single-trunked palms. Woodpeckers need suitable live and dead nesting trees; so, these should be protected in restorations, parks, and gardens, especially royal, sabal, and coconut palms.

Jamaican woodpecker.

They may also use supplemental nest boxes. Woodpeckers tend to do well after hurricanes, which produce dead tree limbs, standing dead tree trunks, and wood-boring insects.

Falcons (*Falco*): American kestrel (*Falco sparverius*) (cernícalo, grigri fran, killy killy) is a permanent resident from The Bahamas into the Lesser Antilles, supplemented by wintering birds in the north. They frequent towns, gardens, and farm areas having high perch sites. Kestrels nest in various cavity situations, including topless dead palms, and may use nest boxes. Peregrine falcons (*Falco peregrinus*) (grigri pelren, halcón peregrino) occur in winter and may visit or reside in towns with high perches, pigeons, and bats.

Parakeets (Psittacidae): As noted earlier, most parakeets encountered in the Caribbean today are nonnative. Where present, parakeets, native or not, are highly social and accumulate in screechy flocks that move around towns, farms, parks, and gardens. Parakeets have a broad diet of seeds, fruits, leaves, shoots, flowers, and even cactus. Most food is from a wide range of native and cultivated fruiting plants; they may come to feeders with fruits and nuts. They also eat crops when available. Native parakeets nest in cavities of one sort or another including in termite nests, cliffs, and palm trunks. The brown-throated parakeet (*Eupsittula pertinax*) (Caribbean parakeet, prikichi, West-Indische parkiet) is a primarily South American species also native to the Leeward Antilles and introduced to the Virgin Islands. It readily frequents gardens during its wanderings. The Hispaniolan parakeet (*Psittacara chloroptera*) (perico, perruche) and Cuban parakeet (*Psittacara euops*) (catey, periquito) are now patchily distributed forest and forest-edge species. They come to parks and gardens; the Hispaniolan parakeet is in urban areas.

Amazon parrots (*Amazona*): The 8 indigenous amazons are among the most characteristic and charismatic of Caribbean birds, and among the region's most endangered. Amazons are noticeable for their loud calling in flight and while foraging in trees. They use forests, scrub, public gardens, and natural gardens, tending to remain in pairs and nesting in cavities in trees and palm trunks, and may use wooden utility poles and nest boxes. They can nibble away destructively at most anything, tearing apart leaves, branches, and parts of palms; but they primarily consume fruits, seeds, nuts, flowers, and stems. Favorite plants include black olive, bourreria, clusia, cocoplum, gumbo limbo, hog plum, lignum vitae, palms, pigeon plum, and seagrape, as well as citrus, guava, mango, and papaya.

Several of the more famous species, from Dominica, Saint Vincent, and Saint Lucia, inhabit protected areas of montane rainforest and depend on the conservation and restoration of their rainforest habitat. They are not expected in lowland parks and gardens. The red-necked parrot (*Amazona arausiaca*) (Dominican blue-faced amazon, lesser Dominican amazon, jaco, red-necked amazon) sometimes does come to gardens in Dominica. The yellow-shouldered parrot (*Amazona barbadensis*) (lora) from Bonaire and introduced to Curaçao is an arid habitat specialist obtaining its water from cactus. The yellow-billed parrot (*Amazona collaria*) (Jamaica amazon), black-billed parrot (*Amazona agilis*) from Jamaica, and Cuban parrot (*Amazona leucocephala*) (Bahama parrot, Cayman parrot, cotorra, loro, perico, rose-throated parrot) from The Bahamas, Cayman Islands, and Cuba come to villages, parks, and gardens with fruiting trees. Nearly all Caribbean parrots are seriously threatened by the illegal pet trade; nesting trees are often cut down to get to the chicks which may be sold locally for pets or smuggled out of the country and sold for tens of thousands of dollars. In addition to hunting for the pet trade, Caribbean parrots face long-term habitat loss, storm-related food shortages and forest destruction, and nonnative predators, which led to population declines requiring habitat restoration and population enhancements. Management tools include protected areas, forest management, pride and other education campaigns to raise awareness about the threatened status of native parrots, captive breeding, artificial post-storm feeding, and nest boxes.

Flycatchers and kingbirds (Tyrannidae): Bold and fearless kingbirds are among the more obvious Caribbean birds. There are over 30 species of flycatchers, many of which come to suitable native habitat, public parks, towns, and gardens. Gray kingbirds (*Tyrannus dominicensis*) (chichery, chincherry, fighter, grive d'hiverenage, grote grijze vliegenvanger, petchary, piperi, pipirit, pippiree, piritata, pitigre, pitirre abejero, pitiwick, rainbird, titiri, tyran gris) are common and conspicuous in summer in natural habitats, towns, parks, gardens, as well as more generally. They are noticeable because of their emphatic calls and aggressive tendencies around their nests in parks, gardens, roadside trees, and structures. They winter from Hispaniola southward. The similar loggerhead kingbird (*Tyrannus caudifasciatus*) (clérigo, pipirit chande, pitirre guatîbere) from The Bahamas, Cayman Islands, and Greater Antilles also is common in summer in pine and other forests, open parks, orchards, rural areas, and villages. Six pewee flycatchers (*Contopus*), endemic to one or a few islands, are rather tame and readily observed in parks and gardens. Eight species of migrant or resident myiarchus flycatchers, such as the widely distributed La Sagra's flycatcher (*Myiarchus sagrae*) and the endemic rufous-tailed flycatcher (*Myiarchus validus*) of Jamaica are birds of forests and plantations that also come to and nest in well-wooded parks and gardens. Caribbean elaenia (*Elaenia martinica*) and yellow-bellied elaenia (*Elaenia flavogaster*) use parks, gardens, native scrub, and natural garden edges.

Flycatchers characteristically not only feed on insects in flight, sallying forth from forest-edge perches, but also feed within shrubs or on ground. Insect-rich plants they use include black ironwood, butterfly sage, buttonwood, native figs, gumbo limbo, madeira, matchwood, sennas, and wild tamarind. Some also eat small fruits, and also flowers. Gray kingbirds are one of the Caribbean's primary fruit seed dispersers.

Jamaican becard (*Pachyramphus niger*) (judy (male), housebird, kissidy, mountain dick (female), rickachay, rickatee): A Jamaican endemic, this flycatcher-type bird ranges widely from hills to lowland forest, treed pastures, and into gardens with trees and dense shrubs. It feeds within the cover of the foliage on small fruits such as native ficus, spiders, tree lizards, and insects and also flycatches and hovers.

Vireos (*Vireo*): Sixteen vireo species occur in the Caribbean, some as migrants from North America and others endemic to individual islands. Two species regularly may come to gardens. Black-whiskered vireo (*Vireo altiloquus*) (bastard grieve, bien-te-veo, chouèque, chuck, greenlet, john-chew-it, john-to-whit, juan chivi, julián chiví, monkeybird, oiseau cannne, père gris, petit panache, piade, tchouèque, ti panach pyas-kòlèt, viréo à moustaches) is a common summer resident and breeding bird throughout the Caribbean. Thick-billed vireo (*Vireo crassirostris*) (chick-of-the village, oiseua canne, shear-bark, ti panach gwo bèk, vireo de Bahamas, zwazo kan) primarily occurs in The Bahamas, Turks and Caicos Islands, and Cayman Islands. These species use a variety of forested habitats including gardens having thick shrub and tree plantings. They feed on a variety of insects and small fruits. Most of the other Caribbean vireos are restricted in range or numbers and not easily observed deep within the forest but are noticed by their incessant calling, the origin of some of their common names. Vireos particularly forage for insects and spiders. Plants used include black ironwood, buttonwood, cocoplum, native figs, gumbo limbo, matchwood, seagrape, sennas, soldierwood, and wild tamarind. They also eat small fruits including bayberry, blacktorch, blolly, bourrerias, butterfly sage, clusias, native figs, firebush, gumbo limbo, sages, marlberry, stoppers, trema, and West Indian lilac.

Crows (*Corvus*): The Caribbean's 5 crow species are quite omnivorous eating fruits, invertebrates, lizards, and scavenging, including in human-dominated landscapes of gardens, parks, plantations, and garbage dumps. Cuban (*Corvus nasicus*) and Jamaican crows (*Corvus jamaicensis*), endemic to their Greater Antilles islands and nearby, use public parks and large gardens, especially those with feeding stations. Hispaniolan (*Corvus palmarum*) and Cuban palm-crows (*Corvus minutus*) use farm plots.

Palmchat (*Dulus dominicus*) (cigua palmera, zwazo-palmis): The palmchat is a species and bird family endemic to Hispaniola and the national bird of the Dominican Republic. Noisy birds, they are a common sight in public parks and private gardens. They build huge communal nests in tall palms and similar structures. They also are attracted to flowers and small fruits such as clusias, native figs, gumbo limbo, lancewood, miconias, royal, sabal, and silvertop palms, tremas, and willow bustic, as well as cultivated fruits such as banana, guava, and mango.

Swallows (Hirundinidae): Owing to their obvious perching and insect-catching flight paths, swallows and martins are commonly observed residents or visitors in towns, parks, gardens, and through the countryside. Two closely related Caribbean-nesting species are the Caribbean martin (*Progne dominicensis*) (caraïbische zwaluw, galebird,

Thick-billed vireo in The Bahamas.

golondrina de Iglesias, golondrina grande, hirondelle, hirondelle à ventre blanc, irondel nwa, purple swallow, swallow, white-bellied martin) and Cuban martin (*Progne cryptoleuca*) (golondrina azul). They occur in towns and parks in summer where they nest in cavities, especially in pines and palms but also in structures. They winter in South America. Bahama swallows (*Tachycineta cyaneoviridis*) may build nests in vents, pipes, light fixtures, cell towers, under roofs, and in nest boxes. In the Greater Antilles, cave swallows (*Petrochelidon fulva*) (golondrina de cuevas, hirondelle fauve, irondel falèz, rainbird) roost and nest in cave entrances, bridges, culverts, and buildings. North American-breeding migrant swallows pass though the Caribbean but are rather uncommon except the barn swallow (*Hirundo rustica*) (boerenzwaluw, Christmas bird, golondrina cola de tijera, golondrina de horquilla, golondrina del norte, hirondelle, hirondelle de chemineé, hirondelle roux, hirondelle rustique, irondèl ke long, swallow). Barn swallows are common, sometimes even abundant, in the Caribbean during migration; some remain as winter residents. Migrant and winter-resident purple martins (*Progne subis*) may also be common to abundant in towns, parks, and gardens during migration, especially in Cuba. Swallows and martins are readily noticed either overhead or perched in groups on snags and utility wires.

House wren (*Troglodytes aedon*) (godbird, oiseau bon dieu, rockbird, wallbird): Nesting resident of the Lesser Antilles, house wrens have distinctive populations on 6 islands, such as the "Grenada wren." They inhabit scrub forests and thick-edge habitats, mostly eat insects and spiders, and are comfortable around human structures, so they readily use parks, yards, and gardens and may accept nest boxes.

Grenada race of the house wren.

Gnatcatchers (*Polioptila*): Tame and endearing little gray birds, gnatcatchers flit obviously around garden trees. Blue-gray gnatcatcher (*Polioptila caerulea*) (catbird, chewbird, cottonbird, rabuíta, spain-spain, ti chwichwi) winters widely in the region including feeding in forests, parks, and garden trees. The Cuban gnatcatcher (*Polioptila lembeyei*) (sinsontillo) is an island-endemic with a patchy distribution primarily in scrub habitats.

Gray catbird (*Dumetella carolinensis*) (blue thrasher, blue thrush, zorzal gato, zwazo chat): One of the more common and apparent wintering residents in the northern Caribbean, especially in Cuba, becoming less common further south. It is a bird of thick vegetation and, in addition to scrub and forest, favors gardens with thick natural plantings, where it takes up seasonal residence and defends its winter territories. Wintering catbirds are attracted to water features in parks and gardens. They eat insects and also small native fruits including bourrerias, butterfly sage, native figs, gumbo limbo, lancewood, mastic, marlberry, pigeon plum, spicewood, stoppers, and wild cherry.

Mockingbirds (*Mimus*): Mockingbirds can be one of the great thrills of the Caribbean as their variable songs fill the air. They frequent parks, gardens, towns, and other settled areas. Several species occur in the Caribbean. Northern mockingbird (*Mimus polyglottos*) (English thrasher, nightingale, rosiyol, rossignol, ruiseñor, sinsonte) is a resident throughout The Bahamas to the Greater Antilles. The Bahama mockingbird (*Mimus gundlachii*) (Hill's mockingbird) occurs in The Bahamas, Cuba, and Jamaica. The tropical mockingbird (*Mimus gilvus*) (grive ds savanes, moqueur des savanes, nightingale, peckow, pickaro) is a common resident in the Lesser Antilles. Mockingbirds frequent gardens and settle in areas with cover, high perches, and adjacent open lawns where they forage on the ground. Bahama mockingbirds tend to use more sheltered sites than do northern mockingbirds.

All mockingbirds are primarily insect-eaters. They forage mostly on the ground for insects but also eat fruit and insects in a large list of trees and shrubs. They are not shy about building nests in parks and gardens, which they may defend from passersby.

Thrashers and tremblers (Mimidae): Thrashers are recognizable mockingbird relatives that are quite at home in habitats ranging from rainforests to human-dominated areas, including farms, public parks, resorts, and private gardens. The pearly-eyed thrasher (*Margarops fuscatus*) (black thrasher, grive-corossol, grosse grive, jackbird, Louis Jo, mangobird, moqueur corossol, pawpaw bird, pie voleuse, soursop bird, truche, wall-eyed thrush, zorzal pardo, zwazo-kowosòl) occurs through much of the Caribbean, especially on smaller islands. Several other thrashers are confined to the Lesser Antilles. Thrashers tend to be unshy but usually stay out of sight under the cover of plants, foraging by tossing aside litter searching for their variable diet of fruits, invertebrates, and even smaller birds. Aggressive, the pearly-eyed thrasher outcompetes other birds for nest sites. Native fruits used include bourrerias, candlewood, cecropia, clusias, cocoplum, firebush, Jamaica caper, mastic, miconias, pigeon plum, satinleaf, Sierra and other palms, snowberry, stoppers, tremas, and West Indian lilac and also cultivated fruits including annonas, banana, guava, and mango. They use structures as well as trees for nesting and shelter. The brown trembler (*Cinclocerthia ruficauda*) and gray trembler (*Cinclocerthia gutturalis*) are related, similar-behaving insectivorous birds endemic to the Lesser Antilles.

Thrushes (*Turdus*): Thrushes tend to be quite obvious birds of gardens and parks. The red-legged thrush (*Turdus plumbeus*) (Bahama thrush, chua-chua, cigua calandra, kwèt-kwèt, ouèt-ouèt, zorzal de patas, zorzal real) is widespread regionally. It is genetically diverse having as many as 6 subspecies. It is replaced by the white-chinned thrush (*Turdus*

Red-legged thrush in The Bahamas.

aurantius) (chick-man-chick, hopping dick, two-penny chick) in Jamaica. They are quite obvious birds of gardens, parks, roadsides, disturbed areas, shade-coffee plantations, and other thick vegetation. They feed mostly on the ground on insects and lizards but also on fruit in shrubs and trees, including bayberry, bourrerias, butterfly sage, native figs, gumbo limbo, lancewood, satinleaf, spicewood, stoppers, and West Indian lilac. Four other endemic thrushes and several seasonal migrants may also occur in plantations and gardens with trees and dense shrubs.

Spindalis (*Spindalis*): Four spindalis occur on individual Greater Antilles islands; their previous name, stripe-headed tanager, describes their striking plumage. They come to natural plantings as well as to parks and gardens. They are western spindalis (*Spindalis zena*) from the northern Caribbean; Puerto Rican spindalis (*Spindalis portoricensis*) (reina mora), the national bird of Puerto Rico; Jamaican spindalis (*Spindalis nigricephala*) (mark head); and Hispaniolan spindalis (*Spindalis dominicensis*) (bannann-mi mòn, cigua de la Española, reina mora). Spindalis eat small fruits, flower buds, and insects generally in trees and shrubs. They may also come to feeding trays providing apples or other fruits. Mostly, spindalis are attracted to a stand or garden having a multiplicity of fruiting and flowering trees and shrubs including blolly, stenostomums, stoppers, and, especially, cedar.

Cuban warblers (*Teretistris*): Represent a bird family endemic to Cuba. The yellow-headed warbler (*Teretistris fernandinae*) and Oriente warbler (*Teretistris fornsi*) are relatively common in various habitats where they feed on invertebrates and small reptiles. They prefer dense shrubs, such as in hedgerows and dense native gardens.

Yellow oriole on Bonaire.

Orioles (Icteridae): Orioles are among the more striking of Caribbean garden birds. They generally use forests, scrub, natural gardens, and other dense habitats. The primarily North American Baltimore oriole (*Icterus galbula*) is a winter migrant. The primarily South American Venezuelan troupial (*Icterus icterus*) occurs naturally in the Leeward Antilles and likely was introduced to Puerto Rico and the Virgin Islands. The yellow oriole (*Icterus nigrogularis*) occurs in the Leeward Antilles. Other orioles are endemic to specific islands or island groups such as Martinique,

Montserrat, Saint Lucia, Cuba, and The Bahamas. Cuban orioles (*Icterus melanopis*), (solibio, guainuba) and Bahama orioles (*Icterus northropi*) come to gardens, the latter facing severe conservation threats owing to development and hurricanes. Although having some differences in habitat, orioles generally forage on a wide variety of fruits such as native figs, insects, and other invertebrates, and nectar including from aloe, bananas, and cactus flowers. They will come to densely treed natural gardens, and some will use tray feeders with cut-open oranges or sugar-water feeders with perches.

Blackbirds (Icteridae): Blackbirds are highly adaptable birds that make themselves at home in towns and other human-dominated places. Most readily come to public parks, resorts, and private gardens especially those with leaky seed feeders. They have an opportunistic diet of seeds, fruits, invertebrates, lizards, frogs, and human discards. Blackbirds typically forage on ground, seasonally in mixed-species flocks. Red-winged blackbirds (*Agelaius phoeniceus*), red-shouldered blackbirds (*Agelaius assimilis*) (mayito de ciénaga, totí de ciénaga), and tawny-shouldered blackbirds (*Agelaius humeralis*) (mayito, merle, ti kawouj, totí mayito) are quite bold visiting parks, gardens, and open-air restaurants. The shiny cowbird (*Molothrus bonariensis*) (chamón, tordo, reskiyè) is common and a brood parasite, laying its eggs in the nests of other species. They will come to seed feeders. The red-winged blackbird is a North American migrant and winter resident in the northern Caribbean. The Bahamas also has a resident subspecies. The endangered Jamaican blackbird (*Neospar nigerrimus*) (black banana bird, wild pine sergeant) is generally a high-elevation bird but comes down to lowlands as well. The Cuban blackbird (*Ptiloxena atroviolacea*) (choncholí, totí) is an endemic, abundant, well-known, human-oriented bird. It is attracted to sweet things, including sugar and sugar-water feeders. It also uses seed feeders and other seeds and grains. It is found broadly in farms, cattle pastures, roadsides, parks, and gardens in villages and cities.

Grackles (Icteridae): Grackles are among the more obvious Caribbean birds as they very typically hang out around people. Caribbean indigenous grackles include the Greater Antillean grackle (*Quiscalus niger*) (chango, chichinguaco, chinchilin, ching ching, cling cling, mèl diab, merle diable, mozambique) from the Greater Antilles; Carib grackle (*Quiscalus lugubris*) (bequia sweet, bilbitin, blackbird, cancangan, credit) from the Lesser Antilles; and great-tailed grackle (*Quiscalus mexicanus*) from Jamaica and San Andrés. Grackles are adaptable and locally common having become decidedly human-focused birds. They are quite tame and readily take to village, town, and city life, frequenting public parks and gardens both individually and in loose flocks. They have a diverse diet from insects to food scraps, preferring to feed on the ground and so will frequent garden bird trays and water features.

Warblers (Parulidae): Warblers abound in the Caribbean. There are over 50 species, 14 endemic mostly to single islands, which are of intense conservation interest. Some nest in the Caribbean; some species winter; others migrate through but are dependent on the islands for resting and refueling on their long journeys. Warblers feed mostly on insects they find within the vegetation; some sally out to flycatch; others feed on the ground – or all of the above. They are the ultimate insect-seeking bird so invertebrate-producing native plants are well used by warblers, including black ironwood, buttonwood, citrus and citrus relatives, cocoplum, native figs, gumbo limbo, lancewood, seagrape, sennas, soldierwood, wild coffee, and wild tamarind. Some warblers eat fruits;

Arrowhead warbler in Jamaica.

important native plants include bayberry, black torch, blolly, clusia, fiddlewood, native figs, firebush, gumbo limbo, lancewood, sages, marlberry, stoppers, porterweed, tremas, and West Indian lilac. A few eat nectar, such as from firebush, swampbush, agave, guava, and sugar-water feeders. Endemic warblers, being habitat specialists and range-restricted, are of conservation concern. The Jamaican endemic arrowhead warbler (*Setophaga pharetra*) ranges into lowland habitats in nonbreeding season. Bahama yellowthroats (*Geothlypis rostrata*) come to parks and gardens. An exception to habitat-restricted ranges is the distinctive, near endemic, nonmigratory subspecies of the yellow warbler (*Setophaga petechia*), which has at least 11 subspecies in the islands, all with variation in plumage. Both resident yellow warblers and wintering birds are habitat generalists, including using parks and gardens, including in cities.

Migrating and wintering warblers tend to use scrub habitats, parks, gardens, and other habitats as they pass through. Cape May warblers (*Setophaga tigrina*) frequent gardens with nectar-bearing winter flowers and sugar-water feeders. Yellow-rumped warblers (*Setophaga coronata*) use thickly planted gardens. Yellow-throated warblers (*Setophaga dominica*), prairie warblers (*Setophaga discolor*), and American redstarts (*Setophaga ruticilla*) also all come to parks and gardens. Abundant palm (*Setophaga palmarum*), black-throated blue (*Setophaga caerulescens*), northern parula (*Setophaga americana*), and Cape May warblers may come to sugar water in a garden. Northern water thrushes (*Parkesia noveboracensis*), common yellowthroats (*Geothlypis trichas*), and many other warblers are attracted to water features, including streams, bird baths, and water drips. One thing that these warblers have in common is that plantings of closely spaced native trees and shrubs, especially those that are common in the scrub and secondary growth, are the best attractants.

Cardinals, euphonias, and tanagers: Four extraordinarily plumaged, endemic euphonias are generally forest birds ranging from mountains and shade-coffee plantations to lowland dry scrub. Endemic to their islands, they are Puerto Rican euphonia (*Chlorophonia sclateri*) (canario del pais, jilguerillo), Hispaniolan euphonia (*Chlorophonia musica*) (Christmas bird, louidò, mistletoe bird, onza de oro), Lesser Antillean euphonia (*Chlorophonia flavifrons*) (avant-noël, organiste Louis d'or, perruche, roy bois), and Jamaican euphonia (*Euphonia jamaica*) (chochoquit, short-mouthed bluequit). Favoring mistletoe fruit, they also eat fruit of cecropia, native figs, miconias, guavas, and bananas. They may occasionally come to well-shaded gardens near forests. Summer tanagers (*Piranga rubra*) (cardinal) and scarlet tanagers (*Piranga olivacea*) (cardinal alinegro, scharlakenrode tanagra) are migrant and wintering birds that specialize in catching wasps. They are not overly common in the Caribbean but will come to thickly vegetated parks and gardens.

Buntings (*Passerina*): Indigo buntings (*Passerina cyanea*) (azulejo), painted buntings (*Passerina ciris*) (miraposa, verdón), and blue grosbeaks (*Passerina caerulea*) (azulejón, azulejo, azulejo real) are attractive garden birds that migrate through or winter in the Caribbean. Buntings are primarily seedeaters and will come to fields, rice paddies, roadsides, grassy parks, and gardens with grasses or seed feeders with dense shrubs nearby. All 3 buntings and many other species are extremely popular as caged birds in Cuba, and thousands are captured illegally for the pet trade every year.

Hispaniolan tanagers: Representing endemic species and an endemic bird family in Hispaniola, palm-tanagers occupy a wide range of habitats including rural areas, public parks, and private and even urban gardens. They seek out trees and tall palms. The black-crowned palm-tanager (*Phaenicophilus palmarum*) (cuatro ojos) and gray-crowned palm-tanager (*Phaenicophilus poliocephalus*) (cuatro ojos, quatre-yeux) occur in dense, mostly montane, understory habitat and forage on a diverse diet of seeds, fruits, and insects and will come to trees and shrubs in appropriately planted gardens. Among the plants they use are blolly, cocoplum, firebush, guava, heliconia, lignum vitae, matchwood, miconia, soldierwood, stoppers, West Indian lilac, and white alling. The Hispaniolan highland-tanager (*Xenoligea montana*) (ciguita aliblanca, ciguita cubera, petit chitte, petit quartre-yeux, white-winged ground-warbler, white-winged warbler) is highly endangered, but the green-tailed ground-tanager (*Microligea palustris*) (ciguita coliverde, green-tailed warbler, siguita, ti chit lasel) can occur at low-elevation gardens.

Bananaquit (*Coereba flaveola*) (bananabird, beenybird, black see-see, cigüita, falle jaune, kit, reinita, see-see bird, sicrié cage, sucrier, sucrier à pointrine jaune, sucrier à ventre jaune, sugarbird, suikerdiefje, teasy, yellowbreast): Being one of the Caribbean's most iconic and charismatic garden birds, bananaquits are found throughout the region except in mainland Cuba. Over 15 recognized subspecies show plumage differences. They are quite comfortable in a park or garden setting and will create obvious, multiple, globular nests through the garden. This nectar-feeding specialist uses some of same plants as hummingbirds but cannot hover and so needs plants it can perch on or at least hang on to briefly. Bananaquits pierce soft fruits and flower bases to lap up juice or nectar and also eat small insects. Among the many flowers that attract bananaquits are agave, aloe, banana, bourrerias, cactus, cinnamon, dogwood, firebush, Geiger trees, guava, heliconia, hibiscus, Jamaica caper, sages, miconia, passionvines, porterweeds, swampbushes, tabebuias, trema, West Indian lilac, white alling, and wild dilly. They are attracted by sweet – they

Black-crowned palm-tanager.

Orangequit in Jamaica.

may drink the juices from fresh fruit and will readily come to bird trays with sweet offerings such as bananas, open oranges, papaya, or cut grapes. Bananaquits come to sugar water in hummingbird feeders, so long as they can perch. They also will eat solid sugar, which is nutritionally inappropriate for the bird.

Orangequit (*Euneornis campestris*) (blue badas, blue baize, long-mouth bluequit): This Jamaican endemic (species and genus) occurs in lowland to midlevel forests and open woodlands as well as in appropriate gardens. They feed broadly on small fruits such as cecropia and native figs and also on flowers, nectar, sapsucker sap, invertebrates, and seeds. They will use degraded habitat, shade-coffee plantations, and gardens with fruiting plants and feeders with fruits such as oranges, bananas, and sugar water.

Bullfinches (*Melopyrrha*, *Loxigilla*): Caribbean bullfinches eat seeds and small hard fruits, but they are related to tanagers and so also eat fruits, insects, spiders, and nectar. They are birds of the edges of diverse wooded habitats including forests, scrub, tree plantations, and degraded forests and generally have acclimated well to humans and occur in towns, parks, and gardens. There are 6 species called bullfinches in the region. The Greater Antillean bullfinch (*Melopyrrha violacea*) (black sparrow, cotton-tree sparrow, gallito, gallito prieto, jack sparrow, père-noire, petit-croq, police bird, tchitchi gwo bèk, ti kòk) and the Lesser Antillean bullfinch (*Loxigilla noctis*) (red breast, robin, semillero gorjirrojo, sporophile petit-coq) are the most widely distributed. Endemic bullfinch species occur on Puerto Rico, Cuba, Cayman Islands, and Barbados. Hurricanes along with habitat reductions and severe hunting as cagebirds have been contributing causes of population declines. Lone bullfinches in cages are common sights. Bullfinches on some, but not all, islands will visit gardens with seed or fruit feeders and even enter houses.

Grassquits (*Melanospiza, Phonipara, Tiaris*): Several species of small, ground-feeding, seed-eating birds commonly called grassquits occur in the Caribbean. In addition to seeds, they also eat small fruits and buds. Grassquits use diverse feeding sites including golf courses, roadsides, public parks, resorts, and private gardens with grass. Natural garden plants visited include beach creeper, cactus, porterweed, Spanish needle, and other composite flowers. They often nest in spiny plants, so cactuses are particularly valuable in a park or garden setting. In addition, diverse lawns, grasses left to go to seed, and seed feeders are enhancements. The black-faced grassquit (*Melanospiza bicolor*) (barbito, black sparrow, cheechee bird, cici-zèb, gorrión nego, grass sparrow, groundquit, juana maruca, mangeur d'herbe, sinbird, sisi zèb, sparrow, sporophile à face noire, tobacco bird, tobacco seed, tomeguín prieto, zwartkopgrasvink) is widespread through the Caribbean with 8 recognized subspecies. The Cuban grassquit (*Phonipara canora*) (senserenico, tomeguín del pinar), naturally endemic to Cuba although introduced to The Bahamas, occurs in semiarid

Cuban grassquit.

habitats and is a popular cage bird. The yellow-shouldered grassquit (*Loxipasser anoxanthus*) (yellowback, yellow-backed finch) is a Jamaican endemic. The yellow-faced grassquit (*Tiaris olivacea*) (cigüita de hierba, gorrión barba amarilla, grassbird, petit z'herbes, squit, tomeguín de la tierra, viudito, zèbable) occurs through the Greater Antilles and uses tall grass, including in gardens. Grassquits adaptably tend to go where there is grass seed.

Lesser Antillean tanager (*Stilpnia cucullata*) (ci-ci corossol, dos-bleu, golden tanager, pawpaw bird, prince bird, soursop bird): These small, interestingly colored forest birds from Grenada and Saint Vincent use scrub and disturbed areas including degraded former forest, parks, and gardens and are attracted to gardens with fruiting trees and shrubs.

6.3 Reptiles

Red-footed tortoise (*Geochelone carbonaria*): Primarily a species of northern South America, this land tortoise occurs also on some of the Lesser Antilles, Puerto Rico, and the Virgin Islands. The origin of Caribbean island populations is unclear; but tortoises disperse naturally over water and could have been brought by Indigenous People. They also are moved around; tortoises encountered are often adopted for a while as pets. Collecting for the pet trade remains a conservation challenge. Caribbean populations broadly use forest, scrub, and beach habitats. They are tame, curious, willing to accept handouts and will come to and settle into a park, neighborhood, or natural garden setting.

Iguanas (Iguanidae): Symbolic of American tropics. Green iguanas (*Iguana iguana*) (garrobo, guana, iguana, iguana verde, iguana vet, iguane commun, ministro) occur naturally in Grenada, Aruba, Curaçao, Bonaire, Granada, the Grenadines, Montserrat, Saba, Saint Lucia, and Saint Vincent. Four subspecies of the green iguana are recognized, 3 from the Lesser Antilles including the Grenadines horned iguana (pink rhino iguana) of

Red-footed tortoise in the Grenadines.

Grenadines horned iguana, a green iguana subspecies endemic to the Lesser Antilles.

Saint Vincent and the Grenadines and Grenada. The Lesser Antillean iguana (*Iguana delicatissima*) (Lesser Antilles green iguana, West Indian iguana) is indigenous to the northern Lesser Antilles. Green iguanas are mostly vegetarian, arboreal, and semiaquatic; they are comfortable eating their way through a garden's flowers, fruits, and foliage. Where native, green iguana populations are part of the natural environment. Where recently invaded, they overpopulate and threaten island-indigenous iguana populations of both species through hybridization, which is already occurring on some islands. Ten species of rock iguanas (*Cyclura*), each endemic to one or a few islands, are threatened due to their small, isolated populations, collecting, and people disturbing and feeding them. On inhabited islands, rock iguanas can range into gardens. Providing fruit-bearing native shrubs is beneficial.

Curly-tailed lizards (Leiocephalidae): Among the most distinctive and

Curly-tailed lizard in The Bahamas.

personable of Caribbean wildlife with tails endearingly curling up over their back and operating in an unafraid manner, curly-tailed lizards may be seen scrambling across parks, gardens, and parking lots. Nearly 30 species are indigenous to Lucayan and Cayman Islands, Cuba, and Hispaniola. The saw-scaled curly-tailed lizard (*Leiocephalus carinatus*) alone has 13 recognized subspecies. Genetic research is likely to reveal more full species on individual islands. They have broad habitat tolerances from dry coastal dunes to parking lots and are at home in gardens and public parks. Individuals can become quite tame, even brazen, and accustomed to come for food offerings. Rock-pile hideaways and sunning spots are garden enhancements.

Tree lizards (*Anolis*) (anole, anolis, lagarto, salamander): Tree lizards or anoles are ubiquitous small lizards of the Caribbean – abundant, obvious, and occurring nearly everywhere.[22–24] There are over 150 extant species, nearly all endemic to one or few islands. Where there is a single species on an island, it can be variable among diverse habitats. Given the paucity of competing species in the Caribbean, they can be quite abundant and widely distributed in different habitats including native forests and scrub, towns, parks, gardens, and other human-dominated situations. They fill roles that on continents are played by other animals such as insectivorous birds, which are relatively uncommon as breeding species in the Caribbean. Islands have species that are specialists for living on

Jamaican giant anole.

Antiguan anole.

the ground, on tree trunks, or on both the ground and trunks; others live in the canopy, still others on twigs; a few are nearly aquatic. They readily hitch rides from one island to the next, so species have come to be introduced beyond their native ranges. Originally endemic species include the Jamaican giant anole (*Anolis garmani*), also introduced to Florida, and Antiguan anole (*Anolis leachii*), also introduced to Bermuda. Among widespread, noticeable, park- and garden-dwelling species are the Puerto Rican crested anole (*Anolis cristatellus*) and Cuban brown anole (*Anolis sagrei*). Both are generalists that have expanded well beyond their native ranges. Cats and mongooses are threats to range-restricted species, and unfortunately, a cultural aversion to anoles (street duppies, African duppies, greenguanas) persists on some islands.

Geckos (Gekkota): Geckos are among the more common animals supported by Caribbean parks and gardens. Hidden during day, they become obvious at night near lights. Over 100 gecko species are native to the Caribbean, and a few others are introduced. Dwarf geckos (*Sphaerodactylus*) (sphaeros, ground geckos) alone number 86 species, most endemic to their island or even part of an island. Species such as the Jamaican stippled sphaero (*Sphaerodactylus argus*) occur in the litter and under logs in natural areas, natural gardens, and around structures. Turnip-tailed geckos (*Thecadactylus rapicauda*) (largarrabo, tuqueque, tuteca) and woodslaves (*Aristelliger praesignis*) (croaking lizard, Jamaican croaking gecko) are found in gardens. Geckos disperse in human cargo and so have been introduced widely outside their native range. Ocellated gecko (*Sphaerodactylus argus*) is a native house gecko from Jamaica and Cuba and now occurs in The Bahamas. Common nonnative species include Brook's house gecko (*Hemidactylus angulatus*) from Africa; Southeast Asian house gecko (*Hemidactylus garnotii*); and African tropical house gecko (*Hemidactylus mabouia*). House geckos are generally considered to bring good luck. Geckos eat insects and other small invertebrates spending the day under rocks in the garden, in trees, within shrubs, on bark, behind shutters, or in the house. Nonnative geckos tend to be especially obvious around houses and gardens. Enhancements include leaving fallen limbs, rocks, and pots around the garden for them to hide in during the day. They are attracted to lights at night.

Ground lizards (*Ameiva, Gymnophthalmus*): Also called ground runners and jungle runners as they can hardly be caught, they number over 20 species in the Caribbean. These large, ground-dwelling lizards are quite at home in botanic gardens, public parks, gardens, and yards. They forage obviously, rooting through leaf litter for fruits, invertebrates, and other lizards. The primarily South American giant ameiva (*Ameiva ameiva/praesignis*) and Antillean ameiva (*Ameiva tobagana*) occur in the Lesser Antilles; and there are several species in the Greater Antilles. The related tegus are omnivorous, egg-eating specialists. Underwood's spectacled tegu (*Gymnophthalmus underwoodi*) is a long-tailed, unisexual species from Antigua, Barbados, Barbuda, Guadeloupe, Martinique, and Saint Vincent and is invasive on Saba and Sint Eustatius. It is a terrestrial lizard occurring in open, sunny, particularly grassy areas of parks and gardens. Ground lizards require heat; they are active during the hottest part of the day and bask on heated rocks and pavement. They forage in litter for fruits, flowers, carrion, invertebrates, small lizards, and eggs. Enhancements include providing PVC pipes as artificial burrows and sunny basking rocks.

Pair of giant ameivas on a garden sidewalk.

Skinks (Scincidae): Mostly ground-dwelling lizards. They are at home in parks and gardens as well as natural areas. When not foraging, they shelter under rock piles, rock walls, garden debris, and buildings. Skinks are active during the day, foraging noticeably in the leaf litter, under shrubs, and along buildings. Overall, there are about 60 skink species in the region, all but a few being endemic to a single island. Antillean four-lined skinks (*Spondylurus*), endemic to the Caribbean, are unique among reptiles being placental live bearers, similar to mammals. Modern genetic tools have identified at least 10 extant species. Two-lined skinks (*Mabuya*) occur on Dominica and Guadeloupe. Many species are endangered due to their highly restricted range and susceptibility to predation by mongooses and cats. Where they occur, Caribbean skinks can be an indicator species for protecting and restoring natural habitat.

Snakes (Serpentes): The Caribbean is a hotspot of snake diversification, including 26 species of dwarf boas (*Tropidophus*) and 41 species of tiny fossorial blind snakes (*Typhlops*). As many as 13 boa species (*Boa*) occur in the region, such as the range-restricted Hispaniolan boa (*Epicrates striatus*) of Hispaniola and The Bahamas (where it is endangered) and the widespread boa constrictor (*Boa constrictor*) which natively ranges into Dominica and Saint Lucia. Tiny blind snakes, worm snakes, and thread snakes (*Leptotyphlopidae, Typhlopidae*) and the nonnative Indian Brahminy blind snake (*Indotyphlops braminus*) are hardly ever seen but live in forests, public parks, and gardens buried in the soil or under logs, leaves, stones, mulch, or debris. Racers (of several genera) (koulèv, serpientes corridor) are likely the snakes most commonly encountered. Endemic on individual islands, they are active, fast, and noticeable searching for insects and lizards during the day in public parks and gardens, and defending themselves. The Antiguan racer (*Alsophis antiguae*) is one of the rarest snakes in the world. Red-bellied racer (*Alsophis rufiventris*) is a small racer in the Lesser Antilles. The brown racer (*Cubophis vudii*) from The Bahamas, Cayman, and Cuba is a fast-moving predator on frogs and lizards, including in parks and gardens. Snakes depend on thick vegetation for protection from people, cats, and mongoose. Caribbean snakes have fared badly from mongoose predation, which has devastated some populations, and also from people predation. As elsewhere, snakes are incorporated into Caribbean folklore and African-Caribbean belief systems and tend not to be among the most favored wildlife. Many do well in parks and gardens, if allowed; but a traditional greeting for a snake is to kill it with a cutlass and be proud of it, which is not helpful for their conservation.

6.4 Amphibians

Rain and southern frogs (Eleutherodactylidae, Leptodactylidae): Hundreds of frog species dot the Caribbean landscape, some highly restricted in range and habitat, others widespread and garden favorites. The smallish rain frogs tend to take well to parks, natural gardens, and greenhouses where they tuck into shrubs or hide under litter, within debris, and in pots. As a result, they can be quite urban and readily hitch rides as their pots and plants are transported elsewhere. The greenhouse frog (*Eleutherodactylus planirostris*) (rana-ladrona de invernadero) occurs naturally in The Bahamas, Caymans, and Cuba but it has been introduced elsewhere including Jamaica. Coqui (*Eleutherodactylus coqui*) (Caribbean tree frog, Puerto Rican frog) can be exceptionally abundant. It is a cultural

Lesser Antilles whistling frog.

symbol in Puerto Rico but an invasive species outside its native range. Lesser Antilles whistling frog (*Eleutherodactylus johnstonei*) (Antilles coqui, Barbados whistling frog, Johnstone's whistling frog, Montserrat whistling frog) is naturally a Caribbean endemic, but now it is widely introduced including to Bermuda as far back as the 1880s. Many rain frogs are small, and some are tiny, less than 1 cm. But the critically endangered giant ditch frog (*Leptodactylus fallax*) (crapaud, mountain chicken) can weigh a kilogram. This and other frogs are at existential risk to the invasive fungal disease, chytridiomycosis, requiring extensive ex situ conservation action. Frogs are mostly active nocturnally foraging for insects, moths, snails, and smaller frogs and call incessantly in breeding season. Rain frogs lay eggs in damp soil of gardens and under rock as they do not require water. Providing hiding places and water features enhances use of a natural garden by frogs. Their nighttime chorus is one of the great sounds of a Caribbean private or resort garden.

Tree frogs (*Osteopilus*): Something of a conservation enigma in the Caribbean. The Cuban tree frog (*Osteopilus septentrionalis*) is native in The Bahamas, Cuba, and Cayman Islands but is widely introduced through the Virgin Islands, Lesser Antilles, and elsewhere where it predates on other indigenous frogs and smaller vertebrates. Red-snouted tree frog (*Scinax ruber*) has been introduced and is invasive on Lesser Antilles islands, Puerto Rico, and Martinique. Tree frogs tend to be garden frogs, so gardens can provide habitat for indigenous species. They spend the day sheltering in herbaceous plants and feed at night, sometimes being attracted to night lighting. Some will use bird boxes and other artificial sites provided in a park or garden. Conservation concern depends on the native status of a species.

6.5 Mammals

Southern opossum (*Didelphis marsupialis*): Opossums may be native to several of the Lesser Antilles having been brought by Indigenous People, or they may have arrived later. Where present, an opossum may be a common garden animal, either sleeping during the day in a secure spot or seen on its nighttime foraging rounds. They will make use of large wildlife boxes and can become quite tame.

Agoutis (*Dasyprocta*): Originally from South America, the Brazilian agouti (*Dasyprocta leporina*) (red-rumped agouti, golden-rumped agouti, orange-rumped agouti) was brought into the Caribbean by Indigenous Peoples millennia ago, and so this is considered to be native in the Lesser Antilles. The related nonnative Central American agouti (*Dasyprocta punctata*) (agutí, ñeque) has been introduced to Cuba, Cayman Islands, and Virgin Islands. Wary, impressively jumpy, and fast to run when disturbed, they can become tame when protected in public parks, resorts, and large private gardens. They prefer to be near water and feed in groups during the day on seeds, fruits, leaves, roots, and insect larvae. How they are welcomed and managed in a restoration or natural garden depends on their native status in a location and their impact on plants.

Hutias (Capromyinae): Endemic to the Caribbean, over 20 species of these rodents have been described, although only 10 range-restricted species remain, mostly in Cuba. The Bahamian hutia (*Geocapromys ingrahami*) is naturally endemic to a few small Bahamian islands. They were eaten by the region's Indigenous People and are hunted, eaten,

Bahamian hutia on Warderick Wells, The Bahamas.

and kept as pets in Cuba. They are wary, but where abundant and protected, they will use park and garden settings, such as at Guantánamo Bay Naval Base in Cuba. Fierce vegetarians with a specialized plant-processing digestive tract, they can devastate an edible garden. Nonetheless, where found, they are of conservation importance.

Bats (Chiroptera): Unlike indigenous terrestrial mammals in the Caribbean, bats are abundant and diverse, occurring in native forests as well as towns, parks, and gardens with species-appropriate features.[54] Nectar-seeking species are among the most important pollinators for some plants, including cactuses and agave as well as mango and avocado. Others eat insects, fruits, or fish. The Cuban flower bat (*Phyllonycteris poeyi*) is a locally abundant species that feeds directly on flower pollen from a wide range of plants. Some of interest that might come to natural plantings in gardens and parks to feed or roost include greater bulldog bat (*Noctilio leporinus*); Jamaican fruit-eating bat (*Artibeus jamaicensis*); Antillean fruit-eating bat (*Brahyphylla cavernarium*); chestnut mastiff bat (*Eumops ferox*) (fierce bonneted bat); and big brown bat (*Eptesicus fuscus*). Fruit-eating and other bats roost in tree cavities, among leaves including palm leaves, under bark, and in caves and cave-substitutes. Fruits eaten by bats include bananas, dilly, figs, guava, mango, matchwood, palms, trumpet trees, and white alling. At least 13 Caribbean bat species are threatened, in part owing to loss of tree and cave roost sites. Some species can be supported by increasing the number of roosting and nesting sites through restoring large native trees to the garden and not cutting dead leaves off palms. Other species use structures; an enhancement is installing bat houses as appropriate for the species that might be nearby. Such enhancements can also be done in public gardens and on public buildings. Bats over the garden are fascinating as they zip noiselessly through the plantings at dusk, gathering flying insects.

6.6 Invertebrates

Snails (Gastropoda): The Caribbean is a land of snails, although unappreciated by most people. There are 700 Caribbean species in the family Annulariidae alone – an entire family of snails endemic to the Caribbean. Jamaica has over 500 snail species, with 96% endemic. A small island like Dominica has over 40 species of land snails and slugs, with 16 endemic. The taxonomy and distribution of Caribbean snails are under-researched; few have common names; and only experts might identify most specimens. Snails and slugs are ecologically important detritivores and herbivores in most habitats including natural areas, parks, and gardens. Terrestrial snails are generally under cover during the day, and they emerge at night to forage on plant leaves, stems, roots, decaying matter, and fungi. Tree snails (*Polymita, Liguus, Hemitrochus, Plagioptucha*) are Caribbean treasures. They feed on algae and lichens on trunks and branches of native forest trees; 84 tree species are recorded to be host plants for Caribbean tree snails. Painted snails (*Polymita*) of eastern Cuba and liguus (*Liguus*) of Cuba and Hispaniola have a multitude of color forms. The Cuban kite (*Chondrohierax wilsonii*) specializes in eating tree snails. The coastal peanut snail (*Cerion*) is one of the most abundant land snails on many islands; over 100 species have been recognized. Slugs also are common, notably including the large leather-leaf slugs such as the Cuban slug (*Veronicella cubensis*) and Jamaican pancake slug (*Veronicella sloanii*), which is now also in the Lesser Antilles and considered a voracious pest outside its range. Other nonnative species have been introduced by various means, mostly as stowaways. The most famous invader is the giant African snail that is found in Cuba, Martinique, and Sint Eustatius and pops up elsewhere. There are thousands of species of land snails in the Caribbean, over 1,300 in Cuba alone. Most native snails encountered will be endemic, a critical part of their ecosystem, and do no damage and indeed benefit natural gardens and other restorations. If not causing trouble, snails and slugs may be accommodated in restorations, parks, and gardens for their ecosystem services and as food for other animals. Food-plant pests are another matter, but there are environmental-friendly ways of dealing with them.

Crabs (Decapoda): In the Caribbean, terrestrial crabs occupy ecological niches held by other species on the mainland. Wide-ranging land species include the blue land crab (*Cardisoma guanhumi*) (cangrejo, giant land crab, juey), purple land crab (*Gecarcinus ruricola*) (zombie crab), and West Indian hermit crab (*Coenobita clypeatus*) (purple pincher crab, soldier crab, tree crab, and other less publishable names when pinched – claws are called pinchers). Other interesting species occur throughout the region, especially on Jamaica, such as bromeliad crabs (*Metopaulias depressus*) and freshwater crabs (*Sesarma*). Land crabs have local names, often based on the local shell color. They serve as important detritivores and herbivores within island ecosystems consuming leaves, fruits, insects, and dead animals. They are also good to eat and are well used in Caribbean cuisines as boiled, shredded, curried crab meat, crab and callaloo soup, crab and dumplings, crabs and rice, and crabs and other such. Nighttime land crab catching is a cultural ritual and commercial undertaking on some islands where they are sold roadside and at market. Towns pride themselves on their crabs. Over-hunting crabs on some islands is a conservation concern.

Statue honoring land crabs in Playa Larga, Cuba.

Terrestrial crabs' populations and natural habitats are decreasing, which is a cultural issue as well as an ecological one. They will use restored areas, parks, and natural gardens where their scavenging is useful, but they may help themselves to garden production, which may be hard to combat – caging plants being the best approach. Land hermit crabs are not so destructive in a garden as are blue land crabs. Blue crabs are burrowers and also amazing climbers in season and hard to repel with the even the most complete of tall barriers. Blue land crabs tend to be coastal. Purple land crabs, hermit crabs, and other species range far inland and up mountains. Seasonal spawning migrations of land crabs called corridas, or tire-eaters, not only serve to ecologically connect the land to the ocean but also pose risks from people collecting or running over them along the way. On most islands, corridas are becoming less abundant. Assuring the retention or creation of natural habitats and protected areas, particularly along the coast, and management of land crab hunting and hermit crabs market harvest deserve conservation attention.

Spiders (Araneae): Spiders are highly diverse throughout the Caribbean. Cuba has over 600 known species. Caribbean spiders range from hardly visible to overly impressive tarantulas. Many species are widely dispersed over the region and beyond because spiderlings travel far on air currents and storm winds. Golden-silk orb-weavers (*Nephila clavipes*) (banana spider, giant wood spider, golden orb-weaver), silver garden orb-weavers (*Argiope argentata*), and tropical orb-weavers (*Eriophora ravilla*) make large, tough webs highly noticeable in the forest or garden. Their presence indicates a good supply of insects

Silver garden orb-weaver spider.

Julia butterfly.

and a sign of successful restoration. Crab spiders (*Gasteracantha cancriformis*) are widely distributed, variable, and tend to find new natural gardens early. Huntsman (*Heteropoda venatoria*) may have an impressive leg span of up to 10 cm. They live up to the name capturing butterflies, moths, and other insects directly instead of spinning webs. Huntsman commonly live around structures such as garden barns and sheds. Antilles pink-toe tarantulas (*Avicularia versicolor*) are kept as pets. There are so many more spiders through the region. Few are actually dangerous to humans, and all serve as natural small predators.

Butterflies and moths (Lepidoptera): Although the diversity of butterflies and moths in the Caribbean as a whole is not completely known, over 350 of species of butterflies and over 1,700 moths have been identified. Cuba alone has over 1,500 species.[55] Only an expert or serious hobbyist would know a fraction.[3,20,26,29] But there is no real reason for a resource manager, park superintendent, landscaper, or gardener to be able to put a name on every animal that shows up. Some are well known, obvious, abundant in the right habitat, widespread in region, and recognizable, at least to taxonomic group. Having butterflies in a restoration, park, or garden is an indicator of its success in recovering nature.

Swallowtails (Papilioninae): Among the more obvious butterflies of natural areas, parks, and gardens having appropriate host plants and nectar flowers, especially sages, various composites, and tropical milkweed. Over 22 species and many subspecies occur in the Caribbean, with 13 species in Cuba alone. Papilio species, such as dusky swallowtail (*Papilio aristodemus*) (Schaus' swallowtail) and Caribbean swallowtail (*Papilio pelaus*) (Antillean swallowtail), use cultivated and native citrus and relatives as larval plants. Battus species, such as pipevine swallowtail (*Battus polydamas*), use native and nonnative pipevines (*Aristolochia*). Bahaman swallowtail (*Papilio andraemon*), which also uses citrus relatives as its larval hosts, has several subspecies and has colonized Jamaica with the help of hurricanes. The critically endangered Jamaican giant swallowtail (*Papilio homerus*) is the largest butterfly in the Western Hemisphere.

Sulphurs (Coliadinae): Many species of sulphur butterflies occur throughout the Caribbean including cloudless sulphur (*Phoebis sennae*), orange-barred sulphur (*Phoebis philea*), large orange sulphur (*Phoebis agarithe*), statira sulphur (*Aphrissa statira*), and barred sulphur (*Eurema daira*). Some are highly migratory and can be seen flying over the water on their seasonal migrations. Sulphurs also are common butterflies of parks and gardens providing appropriate plants. Larval host plants are sennas and other legumes. Their nectar plants tend to have flowers with long tubes, such as bloodberry, bougainvillea, and sages.

Hairstreaks (Lycaenidae): There are many hairstreak butterflies to be found in the Caribbean including species that are endemic to single Caribbean islands. Some include Caribbean scrub-hairstreak (*Strymon acis*) (Bartram's scrub hairstreak), atala hairstreak (*Eumaeus atala*), silver-banded hairstreak (*Chlorostrymon simaethis*), Antillean hairstreak (*Chlorostrymon maesites*), and fulvous hairstreak (*Electrostrymon angelia*). Hairstreaks as a whole use a wide variety of host plants. Some caterpillars consume detritus; some are predators; some secrete honeydew and are attended by ants. Nectar plants are similarly broad, including composites and sages. The atala hairstreak of Cuba and The Bahamas specializes in cycads.

Zebra longwing (*Heliconius charithonia*): This is one of the more obvious butterflies through much of the Caribbean from The Bahamas through the Greater Antilles to Montserrat. They can be abundant in parks and gardens with passionvines, both native and nonnative. Adults eat pollen and nectar, especially using firebush, porterweed, sages, Spanish needle, other composite flowers, and strongbark. At night, they retire to communal roosts in trees.

Fritillaries (Argynninae): The silver-spotted flambeau (*Agraulis vanillae*) (Gulf fritillary) is a noticeable, bright orange butterfly found throughout the Caribbean and common in native plantings and gardens. It uses passionvines as larval hosts. Others include Mexican fritillary (*Euptoieta hegesia*) and variegated fritillary (*Euptoieta claudia*). Nectar flowers used by adults include bougainvillea, sages, scorpiontail, porterweeds, tropical sage, and composite flowers such as Spanish needle and tree seaside tansy.

Julia (*Dryas iulia*) (Flambeau, Julia heliconian butterfly): This sensational orange butterfly, common throughout the Caribbean, uses passionvines as larval hosts and is attracted to many sorts of nectar-bearing flowers, particularly sages.

Malachite (*Siproeta stelenes*): It is a widespread, common, sensational-looking, greenish butterfly that uses forest edges, gardens, and orchards. Wild petunias are host plants. Adults have a varied diet including rotting fruits, nectar, and carrion. They are attracted to feeding stations with fruits. At night, they form communal roosts within the garden shrubs.

Monarchs (*Danaus*): Milkweed butterflies are some of the more popular and sought-after garden and nature-center butterflies. The monarch (*Danaus plexippus*) is a large orange and black butterfly found in gardens and other habitats with milkweeds through much of the Caribbean. Milkweed plant species are readily introduced into restorations, parks, and gardens, especially in butterfly gardens, as further discussed in Chapters 4 and 5. Adults feed on a wide variety of nectar sources including firebush, scorpiontail, tropical

Pleurodonte excellens snail from Hispaniola.

sage, and various composites. Individuals from the North American migratory population may pass through north Cuba, but other populations are nonmigratory residents. Related milkweed butterflies include Jamaican monarch (*Danaus cleophile*), tropical queen (*Danaus eresimus*), and queen (*Danaus gilippus*).

Moths (Sphingidae): Of the thousands of moths in the Caribbean, few are identifiable to most people. The black witch (*Ascalapha odorata*) and giant gray sphinx (*Pseudosphinx tetrio*) are exceptions. These are among the more sensational moths attracted to natural plantings, parks, and gardens. The huge black witch has deep cultural meaning as it is associated with misfortune and death – called duppy bat in Jamaica. Larval host plants are legumes such as senna. Adults sip overripe fruits, especially bananas, and so can be attracted to fruit feeders. The sphinx moth's huge caterpillars feed on frangipani and related plants. Both moths make an exceptional presence in a nighttime garden. Many other moth species will also be using natural and specialty gardens.

Other invertebrates: Beyond crabs, snails, spiders, butterflies, and moths, there are many more invertebrates to be expected in restored reserves, parks, and natural gardens. Any such list is long and incomplete but would include multiple species of ants, bees, beetles, bugs, centipedes, house centipedes, dragonflies, flies, grasshoppers, mantis, millipedes, true scorpions, spiders, wasps, whip scorpions, and worms of many sorts. The vast majority of Caribbean invertebrate species no doubt remain to be discovered and classified. Features in gardens and parks can enhance their use by insects and other invertebrates. Of course, the native trees, shrubs, vines, short-statured plants, diverse lawns, and leaf litter are the primary attractors. Mason and leafcutter bee nest-tubes, log and rock piles, and persistent mud puddles can add value. Water features will bring in dragonflies, damselflies, and other aquatic insects. A natural garden is meant to be an insect garden.

Termites and ants are essential to facilitating rapid turnover of organic matter and nutrients in the tropics. There are over 1,200 species of ants known from the Caribbean, including famously annoying species such as crazy ants (*Nylanderia pubens*) and nonnative fire ants (*Solenopsis invicta*). Introduced species were associated with ant plagues in the colonial era. Ants are nearly everywhere in the Caribbean and perform diverse functions in natural areas and gardens. Beyond eating houses, native termites too serve natural functions; their subterranean, mound, and arboreal nests can be a common feature in a natural garden and used by cavity-nesting birds.

Bees are the most important pollinators of Caribbean plants. Nonnative honeybees (*Apis mellifera*) and indigenous stingless bees (*Melipona*) are established in apiaries in gardens and increasingly in environmental centers for the benefit of ecotourism. Carpenter bees (*Xylocopa*), bumble bees (*Bombus*), leafcutting bees (*Megachile*), orchid bees (Euglossini), alkali bees (*Nomia*), and mason bees (*Osmia*) all are important pollinators, some generalized, others plant-species specific. Bees can cover a nectarous inflorescence. Wasps and many hundreds of species of flies similarly are attracted to nectar flowers.

Grasshoppers are among the most recognizable and familiar insects. They are highly diversified in the Caribbean and found in all habitats including forests, scrub, waste areas, and gardens. Grasshoppers include many slant-faced grasshoppers such as the impressive and widespread spotted-winged grasshopper (*Orphulella pelidna*). Caribbean meadow katydids (*Conocephalus cinereus*) occur throughout the Caribbean in diverse habitats including parks and gardens. Caribbean pygmy grasshoppers and pygmy jumping leaves come in a dazzling array of shapes and colors.

There are nearly 40 morphologically diverse species of praying mantis in the Caribbean including species that represent lineages that arrived there 90 million years ago, likely the most ancient lineages of Caribbean animals. There are 25 Caribbean genera. The thick and cryptic grizzled mantid (*Gonatista grisea*) occurs in the Greater Antilles.

Over 2,600 species of native beetles have been described in Cuba alone and 250 on the much smaller island of Barbados.[56] There is no end to the supply of beetles: blister beetles, click-beetles, Hercules beetles, lycid beetles, scarab beetles, and tiger beetles are some of diverse forms. Most can be identified to the sort of beetle, although unlikely to species unless it is rather sensational and well known, such as giant Hercules beetle (*Dynastes hercules*) or Jamaican rhinoceros beetle (*Calypsoryctes dynastoides*). A diversity of beetles is meant to occupy any natural area, whether it is a reserve, park, or garden.

References

1. Raffaele, H. A., J. W. Wiley, O. H. Garrido, A. R. Keith, and J. I. Raffaele. 2020. *Birds of the West Indies*. Princeton: Princeton University Press.
2. Gill, F., and P. Rasmussen, eds. 2023. "IOC World Bird List, v13.1." DOI:10.14344/IOC.ML.13.1
3. Butterflies of Cuba. 2023. "Butterflies of Cuba." www.butterfliesofcuba.com
4. Crother, B. I. 1999. *Caribbean Amphibians and Reptiles*. San Diego: Academic Press.
5. Gannon, M. R. 2005. *Bats of Puerto Rico: An Island Focus and a Caribbean Perspective*. Lubbock: Texas University Press.
6. Gebracht, J., and A. Levesque. 2019. "Complete Checklist of the Birds of the West Indies." www.birdscaribbean.org/wp-content/uploads/2019/10/Complete_WIChecklist_v1-1.pdf
7. Garraway, E., and A. Bailey. 2005. *Butterflies of Jamaica*. London: Macmillan Caribbean Natural History.
8. Garrrido, O. H., and A. Kirkconnell. 2000. *Field Guide to the Birds of Cuba*. Ithaca: Comstock Publishing Associates.
9. Hallett, B. 2006. *Birds of the Bahamas and the Turks & Caicos Islands*. Oxford: Macmillan Caribbean.
10. Kurta, A., A. Rodriguez-Duran, and M. R. Gannon. 2005. *Bats of Puerto Rico: An Island Focus and Caribbean Perspective*. Lubbock: Texas Tech University Press.
11. Kurta, A., and A. Rodriguez-Duran, eds. 2023. *Bats of the West Indies, a Natural History and Field Guide*. Ithaca: Comstock Publishing Associates.
12. Kushlan, J. A., and K. Hines. 2014. *Attracting Birds to South Florida Gardens*. Gainesville: University Press of Florida.
13. Malhotra, A., and R. S. Thorpe. 1999. *Reptiles & Amphibians of the Eastern Caribbean*. Oxford: Macmillan Education.
14. Raffaele, H. A., and J. W. Wiley. 2014. *Wildlife of the Caribbean*. Princeton: Princeton University Press.
15. Raffaele, H., J. Wiley, O. Garrido, A. Keith, and J. Raffaele. 1998. *A Guide to the Birds of the West Indies*. Princeton: Princeton University Press.
16. Reynolds, R. G., R. W. Henderson, L. M. Diaz, T. M. Rodriguez-Cabrera, and A. R. Puente-Rolon. 2023. *Boas of the West Indies: Evolution, Natural History, and Conservation*. Ithaca: Comstock Publishing Associates.

17. Smith, D. A., L. D. Miller, and J. Y. Miller. 1994. *The Butterflies of the West Indies and South Florida*. Oxford: Oxford University Press.
18. Rivero, J. A. 1998. *The Amphibians and Reptiles of Puerto Rico*, 2nd edition. San Juan: Editorial de la Universidad de Puerto Rico.
19. Schwartz, A., and R. W. Henderson. 1991. *Amphibians and Reptiles of the West Indies: Descriptions, Distributions, and Natural History*. Gainesville: University of Florida Press.
20. Riley, N. D. 1975. *Guide to the Butterflies of the West Indies*. Collins: New York.
21. Stiling, P. D. 1999. *Butterflies of the Caribbean*. London: Macmillan Caribbean.
22. White, A. W. 1998. *A Birder's Guide to the Bahama Islands*. Colorado Springs: American Birding Association.
23. Roughgarden, J. 1995. *Anolis Lizards of the Caribbean*. Oxford: Oxford University Press.
24. Schettino, L. R., ed. 1999. *The Iguanid Lizards of Cuba*. Gainesville: University Press of Florida.
25. Losos, J. 2009. *Lizards in an Evolutionary Tree: Ecology and Adaptive Radiation of Anoles*. Berkeley: University of California Press.
26. Green SXM. 2019. "Planting a Caribbean Wildflower and Pollinator Garden." www.greensxm.com/caribbean-wildflower-and-pollinator-gardens
27. Turner, T. W., and V. A. Turland. 2017. *Discovering Jamaican Butterflies and Their Relationships Around the Caribbean*. Safety Harbor: Caribbean Wildlife Publication.
28. Peh, K. S. H., A. Balmford, J. C. Birch, C. Brown, S. H. M. Butchart, J. Daley, J. Dawson, et al. 2015. "Potential Impact of Invasive Alien Species on Ecosystem Services Provided by a Tropical Forested Ecosystem: A Case Study from Montserrat." *Biological Invasions* 17:461–75.
29. Caribbean Invasive Alien Species Network. 2023. "Caribbean Invasives." www.caribbean-invasives.org
30. Scheidt S. N., and S. H. Hurlbert. 2014. "Range Expansion and Population Dynamics of an Invasive Species: The Eurasian Collared-dove (*Streptopelia decaocto*)." *PLOS One* 9(10):e111510. DOI:10.1371/journal.pone.0111510
31. Homan, H. J., R. J. Johnson, J. R. Thiele, and G. M. Linz. 2017. *European Starlings*. Wildlife Damage Management Technical Series. Fort Collins: National Wildlife Research Center. www.aphis.usda.gov/wildlife_damage/reports/Wildlife%20Damage%20Management%20Technical%20Series/European-Starlings-WDM-Technical-Series.pdf
32. Arce-Nazario, J. A., and T. A. Carlo. 2012. "*Iguana iguana* Invasion in Puerto Rico: Facing the Evidence." *Biological Invasions* 14:1981–4.
33. Caribbean Invasives. 2023. "The Invasive Green Iguana." Caribbean Invasive Alien Species Network. www.caribbeaninvasives.org/index.php/2013/01/02/the-invasive-green-iguana-iguana-iguana
34. Shine, R. 2010. "The Ecological Impact of Invasive Cane Toads (*Bufo marinus*) in Australia." *Quarterly Review of Biology* 85:253–91.
35. Wilson, B. S., S. E. Koenig, R. van Veen, E. Miersma, and D. G. Rudolph. 2011. "Cane Toads, a Threat to West Indian Wildlife: Mortality of Jamaican Boas Attributable to Toad Ingestion." *Biological Invasions* 13:55–60.
36. UNEP. 2020. "Caribbean Wrestles with Mischievous Invaders: Monkeys." www.unep.org/news-and-stories/story/caribbean-wrestles-mischievous-invaders-monkeys
37. Quinn, J. H., and D. A. Whisson. 2004. "The Mongoose in the Caribbean: Past Management and Future Challenges." *Proceedings of the Vertebrate Pest Conference* 21:31–6. www.escholarship.org/uc/item/06w9371v
38. Ortega-S., J. A., J. Delgado-Acevedo, J.G. Villarreal-González, R. Borroto-Páez, and R. Tamez-González. 2020. "Wild Pigs in Mexico and the Caribbean." pp. 423–38 in K. C. VerCauteren, J. C. Beasley, S. S. Ditchkoff, J. J. Mayer, G. J. Roloff, and B. K. Strickland, eds.

Invasive Wild Pigs in North America: Ecology, Impacts, and Management. Boca Raton: CRC Press. DOI:10.1201/b22014-18

39. National Park Service. 2003. *Final Environmental Assessment: Sustained Reduction Plan for Nonnative Wild Hogs Within Virgin Islands National Park. St. John, VI.* National Park Service. www.nps.gov/viis/learn/nature/upload/nps_hogs_3002.pdf
40. Borroto-Páez, R. 2009. "Invasive Mammals in Cuba: An Overview." *Biological Invasions* 11:2279–90.
41. Swimming with Pigs. 2023. "Swimming with the Pigs in the Bahamas." www.swimmingwithpigs.com/swimming-pigs-exuma/
42. CARICOM. 2016. "Local Ruminants: Reducing Reliance on Imports." www.caricom.org/local-ruminants-reducing-reliance-on-imports/
43. Hingston, M. 2020. "Ravenous Wild Goats Ruled this Island for Over a Century: Now, It's Being Reborn." *National Geographic.* www.nationalgeographic.com/science/article/raveous-wild-goats-ruled-this-island-for-over-a-century-being-reborn
44 Fleming, G. M., D. N. Ewert, and J. M. Wunderle. 2019. "Response of Early-Successional Bahamian Dry Forest Habitat to Goat Grazing and Implications for Kirtland's Warbler (*Setophaga kirtlandii*) Wintering Ground Management." *Caribbean Journal of Science* 49:157–84. DOI:10.18475/cjos.v49i2.a6
45. Torres-Hernández, G., J. A. Maldonado-Jáquez, L. D. Granados-Rivera, M. Wurzinger, and A. A. Cruz-Tamayo. 2021. "Creole Goats in Latin America and the Caribbean: A Priceless Resource to Ensure the Well-being of Rural Communities." *International Journal of Agricultural Sustainability* 20. DOI:10.1080/14735903.2021.1933361
46 Campbell, E. W. 1991. "The Effects of Introduced Roof Rat on Bird Diversity of Antillean Cays." *Journal of Field Ornithology* 62:343–8.
47. Blackburn, T. M., P. Cassey, R. P. Duncan, K. L. Evans, and K. J. Gaston. 2004. "Avian Extinctions and Mammalian Introductions on Oceanic Islands." *Science* 305:1955–8.
48. Howald, G., C. J. Donlan, J. P. Galvan, J. C. Russell, J. Parkes, A. Samaniego, Y. Wang, et al. 2007. "Invasive Rodent Eradication on Islands." *Conservation Biology* 21:1258–68.
49. Duron, Q., A. B. Shiels, and E. Vidal. 2017. "Control of Invasive Rats on Islands and Priorities for Future Action." *Conservation Biology* 31:761–71. DOI:10.1111/cobi.12885
50. Medina, F. M., E. Bonnaud, E. Vidal, B. R. Tershy, E. S. Zavaleta, C. Josh Donlan, B. S. Keitt, M. Corre, et al. 2011. "A Global Review of the Impacts of Invasive Cats on Island Endangered Vertebrates." *Global Change Biology* 17:3503–10.
51. Mara, P. P. E., and C. Santella. 2016. *Cat Wars: The Devastating Consequences of a Cuddly Killer.* Princeton: Princeton University Press.
52 American Bird Conservancy. 2023. "Cats Indoors." www.abcbirds.org/program/cats-indoors
53. Mead, A. R. 1961. *The Giant African Snail: A Problem in Economic Malacology.* Chicago: University of Chicago Press.
54. USDA. date unknown. *Giant African Snail.* National Invasive Species Information Center. www.invasivespeciesinfo.gov/terrestrial/invertebrates/giant-african-snail
55. Bat Conservation International. 2023. "Bat Houses and Gardens." www.batcon.org/resources/getting-involved/bat-houses
56. Mancina, C. A., and D. D. Cruz Flores, eds. 2017. *Diversidad Biológica de Cuba: Métodos de Inventario, Monitoreo, y Colecciones Biológicas.* Havana: Editorial AMA.

7 Reflections on recovering Caribbean nature

Recovering Caribbean nature depends on re-creating and restoring patches within an islands' landscape in which native plants and the native birds, lizards, frogs, butterflies, and other indigenous animals, along with the rest of their ecosystems, can survive and thrive. While re-creating ecosystems of 1492 will not happen, approaches to it can. Every bit helps, recovering Caribbean nature from a mountain-covering park to securing a small patch of private garden and enhancing their connectivity.

Our considerations in this book are underlined by a theme that the single-most critical path to recovering semblances of the Caribbean's natural environment is through its indigenous plants. Their use can take many forms. Whether considering vast or patchwork ecosystem restorations in protected areas, relandscaping public parks and gardens, establishing private natural gardens, regreening a community, or adding nature-friendly vegetation to commercial sites, regreening resorts, or enhancing public places, native plants provide the basis for how Caribbean nature is recovered. These plants, in the appropriate diversity and abundance for a particular site, create the habitat that birds and other indigenous animals can occupy. Together, indigenous plants and animals and an unseen microcosm of microorganisms create, or re-create, some approach to what the natural ecosystem might have been before humans arrived to change it.

Birds, butterflies, and other animals indigenous to a site will be attracted to the native plant-based habitat that develops there. These animals need the plants to provide food consisting of the insects and other invertebrates that live in and on the plants and in the natural leaf litter under the plants, the fruits and seeds the plants produce, nectar, oils, and pollen in flowers, as well as the flowers, buds, fruits, seeds, and leaves themselves. In this book, we identify some of the many native plants that have flowers, fruits, and seeds specifically structured to attract birds to pollinate the flowers or to disperse seeds. Other native plants depend on bats for pollination or seed dispersal. Beyond butterflies and moths, many Caribbean plants depend on insects for pollination, especially bees but also flies, beetles, and wasps. Animals that do not provide pollination services also use these nectar and pollen sources. Pollinating insects come and go; insects and other invertebrates also live more permanently with the plant's structure. As also catalogued in this book, native plants "host" specific and generalized invertebrates: caterpillars that eat leaves; sap-sucking insects; ants that tend them; termites in dead branches; beetles under the bark; and a detritus-based invertebrate community hidden in branch crotches, nestled within epiphytes, accumulating in palm boots, and living within leaf litter mulch on the ground below. These insects and other invertebrates are a food supply for the predators –

DOI: 10.1201/9781003396321-8

spiders, wasps, mantids, frogs, lizards, and birds. Mobile predators are attracted to and explore the plants, searching for potential prey. The predators too are prey – spiders for wrens, wasps for tanagers, frogs for snakes, lizards for birds, and birds for other birds.

Plants and the structure they create provide animals with cover from their predators, shelter from the elements, and structure that allows places to roost and breed. Butterflies and moths need vegetation to rest in and airy spaces for courtship flights. Tree lizards need their species-specific plant structure such as trunks or branches and need soil for their eggs. Frogs need hiding places and, other than rain frogs, need water for their eggs and tadpoles. Spiders need structure from which to hang webs or soil and litter on the ground in which to burrow. Many more invertebrates need the cover of leaf litter habitat that sprinkles the ground under the canopy or accumulates in the branches. Water is captured by leaves, bromeliads, flowers, and ground features, which become exceptionally important during the annual dry season when natural water sources may become scarce.

In restoring its nature, the Caribbean has challenges that are rather profound, in no small part owing to the history of environmental abuse. In the colonial era and following, nearly all the space that could be farmed was. The forests that covered the islands when Europeans first arrived were burned, roots grubbed out, soil plowed and eroded, and kept that way for centuries. Opportunities for the reestablishment of native vegetation were slim and happenstantial. Subsequent reliance on global economic and agricultural trends further determined the fate of the landscape. More recently, development is taking its toll. Small island nations are wanting for the resources to tackle their environmental issues along with all else that needs to be done.

Distant view of 2 volcanic mountains along the shore of Saint Lucia with the town of Soufrière in the image center, surrounded by tropical forest.

The Caribbean has embraced the international goal of sustainable development – development inherently tied to a sustainable environment. The 2030 Agenda for Sustainable Development was acceded to by all Caribbean nations in 2015; and the 2023 Global Biodiversity Agreement provides bold goals to protect the ecosystems and plant and animal habitats of a large portion of the earth's surface. Similarly, the United Nations Decade on Ecosystem Restoration provides goals to prevent, halt, and reverse the degradation of ecosystems. So, restoration of the terrestrial habitat is a region-wide policy. Most of the nations of the region have agreed to restore 30% of the terrestrial landscape and secure the same percentage in protected areas or similarly managed sites. Some nations have already met the goal. The Caribbean is a global hotspot of biodiversity owing to its indigenous terrestrial biota, its many island-endemic species across taxons, and its history of habitat loss. Important Bird Areas, Key Biodiversity Sites, Zero Extinction Sites, wetlands of international importance, natural heritage sites, biosphere reserves, wildlife corridors, national parks and other protected areas, botanic gardens, and public gardens have been delineated across the Caribbean landscape.

These government-endorsed places provide an essential basic network, or backbone, for meeting goals for a sustainable environment, protected area management, and ecosystem restoration. Beyond reserves, other public lands such as active parks, the land around public buildings, roadsides, medians, town centers, and schools can be landscaped in more naturalistic ways. Publicly owned and managed botanic gardens can be key players. Many that were historically intended to support nonnative plant introduction are now mostly for public use and can be reimagined in part as naturalistic gardens. Public landscaping has a tendency to deteriorate rapidly owing to the cost of and lack of priority for management, particularly if the designs require fertilizing, mowing, trimming, plant replacement, mulching, and irrigation. Money, person power, will, priorities, and loss of the original landscaping concepts from institutional memory soon turn such projects into invasive weed patches. The alternative of natural landscaping public places is cheaper, and easier and carry-on costs are minimal. All this should be desirable to public-land managers.

However essential, the protection and management of public lands are not sufficient. Much of the land of the Caribbean is privately owned, and it is here that rapid and meaningful progress can be made through the commitment of individual private landholders and commercial ventures including, perhaps especially, resorts and tourism facilities. The engagement of non-governmental organizations such as diverse conservation groups, environmental education centers, and private botanic gardens can create awareness and carry on restoration programs. Organic farms, farm-to-table co-ops, backyard gardens, and other small grower landholdings can participate. For the smallholder, the crucial decision is choosing to create a more-natural yard, garden, or larger tract. After that comes implementation. This starts by designing with nature primarily using native plants while designing also for other compatible uses.

Then comes plant selection, which becomes increasingly important and selective as plot size decreases. The diverse Caribbean flora provides many options for plant selection. In this book, we emphasize species occurring and available widely through the region. While these can provide major native-plant structure, each island has its own indigenous and endemic plants to add to restorations. The ecosystem being created, no matter how

Blue-headed quail-dove, an endangered Cuban endemic species, in a private wildlife-friendly garden.

small, adds to the inventory of nature recovery. Importantly, these increasingly function as they are interconnected by corridors of suitable habitat and close stepping-stone plots. The birds, butterflies, and many more less-apparent animals being supported by a restored or regreened plot are a large part of the Caribbean biodiversity that natural sites are protecting.

There are many more practical matters to attend to, from full-scale government-invested ecosystem restoration on reserved areas to individuals in their private gardens. At both ends of the scale, issues are more similar than not: soils or lack of them; water and rainfall; climate and climate change; management practices; staffing; funding and continued funding commitment; hurricanes; demands for multiple use; neighbors; development pressure; available aid and advice; nonprofit organization involvement; and community perceptions. Nature recovery is never easy. Designation, protection, and creative management of reserved areas, large and small, are essential to the overall conservation of national natural resources and heritage. Once identified, the promise of nature recovery remains to be fulfilled and appropriate restoration and management undertaken.

Island conservation is a shared responsibility of government and civil society as a whole. Enabling and supporting local leadership and stakeholder involvement are national and international frameworks. Fortunately, many of the players are in place – community leaders and community-based partnerships, botanic gardens and environmental education centers, natural resource and agricultural agencies, and national nongovernmental organizations. Over the last couple of decades, following the wide recognition of the importance of the Caribbean in world biodiversity conservation, international organizations have engaged with Caribbean governments, local and international NGOs, and citizen stakeholders to evaluate, plan, model, and execute conservation plans and to build local and national professional capacity thereby increasing community support and educational opportunities for conservation action. Given the recognized importance of conserving Caribbean biodiversity, there have developed a number of past and ongoing programs that bring civil society together on behalf of animals and habitat conservation. Saint Lucia's Conservation through Pride Initiative, West Indian Whistling-Duck Program, Caribbean Endemic Bird Festival, International Migratory Bird Day, Americas Flyway Initiative, and Important Bird Areas community-support programs are but a few examples underpinning bird conservation, led by BirdsCaribbean and BirdLife International. Other regional programs include Caribbean Natural Resources Institute, international funders such as the Critical Ecosystem Partnership Fund and Development Bank of Latin America and the Caribbean, and international environmental conventions such as the Global Biodiversity Framework and the United Nations Framework Convention on Climate Change

How might a movement to increasingly recover Caribbean nature move forward? In this book, we suggest that there can be a common vision that begins with a native butterfly-friendly plant in a container and ends seamlessly through connected habitat to large government reserves. This goal of matrices of recovered natural patches joins government and private initiatives. Achieving such a goal depends on the enormous mostly untapped opportunity for recovering nature on the privately owned landscape of private yards and gardens, as well as farms, hotels and resorts, tourist attractions, and golf clubs. Botanic gardens and forestry departments might lead, but this role also falls to the private sector – to plant nurseries, garden centers, landscapers and landscape architects, conservation and restoration practitioners, native plant societies, bird and other conservation organizations, environmental centers, and private natural gardeners. These individual private-sector opportunities are the keepers and providers of the Caribbean's indigenous plant resources and the native animals they support.

Given all the stakeholders who care about Caribbean nature and sustainable development, what might be useful is that communal vision for multi-scale restoration noted earlier supported by a common philosophical approach, practical baseline of understanding, a unifying restoration vocabulary. Recovering nature one plot at a time, as advocated in this book, need not occur in a vacuum. Each of these actions contributes to much larger local, national, and international priorities that together cover the entirety of the biological diversity of the Caribbean. The value and impact of a single home natural garden are not in that plot alone but in its ability to be multiplied by a network of similar initiatives of which it is an integral part, especially to the extent they are connected. No bit of nature recovery is too large or too small. Each agency, organization, business, and individual can do their part to regreen the green Caribbean, island by island, plot by plot, tree by tree, bird by bird.

8 Resources

Many resources exist that provide background and information pertinent to recovering Caribbean nature. Most are published, but increasingly information is available online. They range from technical references to pictorial identification aids.

8.1 Plants and plant communities

Acevedo-Rodríguez, P., and M. T. Strong. 2023. "Flora of the West Indies." Smithsonian Institution. naturalhistory2.si.edu/botany/WestIndies/

Online resource for the catalogue of West Indian seed-bearing plants, background materials, history of plant exploration, and references.

Austin, D. F. 2004. *Florida Ethnobotany*. Boca Raton: CRC Press.

This encyclopedic book summarizes information on the uses and history of more than 800 species of native plants found in Florida. Entries cover information from throughout a plant's range and so are pertinent to the Caribbean as well.

BirdsCaribbean. 2015. "Heritage Plants. Native trees and Plants for Birds and People in the Caribbean." www.birdscaribbean.org/wp-content/uploads/2014/06/Heritage-Plants-Birds Caribbean.pdf

Informative, nontechnical information on using native plants in the Caribbean.

BirdsCaribbean. 2015. "Patrimonio Vegetal. Plantas y árboles nativos para las aves y las personas del Caribe." www.birdscaribbean.org/wp-content/uploads/2014/06/Patrimonio-Vegetal-BirdsCaribbean.pdf

Informative, nontechnical information on using native plants in the Caribbean.

BirdsCaribbean. 2015. "Native Plants for a Bird-Friendly Yard." www.youtube.com/watch?v=IGdue-4cJ6E

Webinar featuring ornithologists/conservationists and native plant experts. Presentations provide information on threats that birds face, how we can help birds by planting plants that are valuable to birds in our yards, 10 plants that are great for migrant and resident birds, and how to get started planting a bird-friendly yard.

DOI: 10.1201/9781003396321-9

Clement, Y., P. Sutherland, P. Moodley and B. Chevannes. 2014. *Herbal Medicine Practices in the Caribbean*. New York: Routledge.
A thorough review of natural products and healing practices in the Caribbean.

Currie, D., J. M. Wunderle, E. Freid, D. N. Ewertm and D. J. Lodge. 2019. *The Natural History of The Bahamas: A Field Guide*. Ithaca: Comstock Publishing Associates.
A guide to selected plants and animals of The Bahamas.

Cutts, W. 2004. *Trees of The Bahamas and Florida*. Oxford: Macmillan Education.
An identification guide using drawings, photographs, and descriptions. Suitable for use more widely in the region.

Gann, G. D., and J. C. Trejo-Torres. 2015–2023. "Plantas de la Isla de Puerto Rico." Delray Beach: Institute for Regional Conservation. www.regionalconservation.org/ircs/database/site/IntroPR.asp
Database with accounts for over 3,000 plant species.

Gann, G. D., C. J. Abbott, C. G. Stocking, K. N. Hines and collaborators. 2005–2023. "Natives For Your Neighborhood." Delray Beach: Institute for Regional Conservation. regionalconservation.org/beta/nfyn/default.asp
An indispensable guide to native plants of South Florida, most of which are from the Caribbean. Helpful species accounts.

Kushlan, J. A., and K. Hines. 2014. *Attracting Birds to South Florida Gardens*. Gainesville: University Press of Florida.
Companion to the present book, describes natural gardening in South Florida, also having broad applicability to the limestone islands of the Caribbean.

Leon Levy Preserve. 2023. "Flora by Common Name." levypreserve.org/flora-by-common-name/
Species accounts of over 400 Caribbean plant species.

Little, Jr., E. L., R. O. Woodbury and F. H. Wadsworth. 1974. *Trees of Puerto Rico and the Virgin Islands*. Agricultural Handbook 449. Washington, DC: U.S. Department of Agriculture, Forest Service.
A technical identification guide to trees of U.S. territories. Pertinent to much of the Caribbean.

Mancina, C. A., and D. D. Cruz Flores, eds. 2017. *Diversidad Biológica de Cuba: Métodos de Inventario, Monitoreo, y Colecciones Biológicas*. Havana: Editorial AMA.
Multiauthor summary and synthesis of Cuban biodiversity incorporating analyses of animal and plant groups.

Nellis, D. W. 1994. *Seashore Plants of South Florida and the Caribbean*. Sarasota: Pineapple Press.
Accounts, identification guide, propagation of widely dispersed shore plants, many of which are recommended for natural plantings in Caribbean.

University of West Indies. 2023. "Plants of the Eastern Caribbean." ecflora.cavehill.uwi.edu/
Database on eastern Caribbean plants applicable more widely.

Quattrocchi, U. 2019–2000. *CRC World Dictionary of Plant Names, Volume IV*. Boca Raton: CRC Press.
A reference providing taxonomic background on over 22,000 genera of plants including scientific, common, and trade names.

Royal Botanic Garden Kew. date unknown. "Plants of the World." powo.science.kew.org/; World Flora Online. 2021. "WFO Plant List." www.worldfloraonline.org/

Related online sources accessing taxonomy, identification, images, distribution, traits, threat status, molecular phylogenies, and uses of over a million species including over 300,000 images of the vascular plants of the world.

Sánchez, C. 2021. "Inventario de los Licófitos y Helechos de Cuba." *Revista del Jardín Botánico Nacional* 42:1–53.

Monographic treatment of a large subset of Caribbean ferns.

Walters, B. B. 2019. *The Greening of Saint Lucia: Economic Development and Environmental Change in the West Indies*. Kingston: The University of West Indies Press.

Study of history of environmental change in Saint Lucia. Applicable to the greater Caribbean, providing insight into the changing role of agriculture on one of the greenest Caribbean islands.

Wilmanowicz, R. 2008–2010. *Bush Medicine in the Bahamas: A Modern Approach*. Germany: CreateSpace.

A densely presented book on medicinal preparations.

8.2 Animals

Acosta, M., L. Mugica, D. Denis, D. Rodríguez, A. Jiménez and O. Torres. 2003. *Aves Comunes en los Humedales de Cuba*. Havana: Universidad de la Habana.

A thorough review of the birds to be found in Cuba's wetlands.

Acosta Cruz, M., L. Mugica Valdes and K. Aguilar Mugica. 2022. *Aves del Jardin Botanico Nacional de Cuba*. Havana: Sello Editorial AMA.

A well-researched and beautifully presented account of the biology, status, and descriptions including photographs of birds found in Havana's botanic garden.

Alliance for Zero Extinction. 2023. "Global AZA Map." zeroextinction.org/site-identification/2023-global-aze-map/

Online map of Alliance for Zero Extinction sites accessible by site, country, or species.

BirdsCaribbean. 2023. "About Us." www.birdscaribbean.org/

Principal regional scientific and conservation organization on the forefront of conservation of native birds and their habitats. The website has comprehensive information. Programs include the *Journal of Caribbean Ornithology*, biennial International Conferences, Caribbean Waterbird Census, Seabird Conservation, Caribbean Landbird Monitoring (including PROALAS surveys, Caribbean Bird Banding Network and the Caribbean Motus Collaboration), Caribbean Birding Trail, and education programs including the Caribbean Endemic Bird Festival, World Migratory Bird Day, West Indian Whistling-Duck and Wetlands Conservation Project, and BirdSleuth Caribbean.

BirdsCaribbean. 2023. "Caribbean Birding Trail." caribbeanbirdingtrail.org/

A network of the best sites on each island for observing birds and otherwise enjoying nature. Includes information on featured birding sites, local guides, and tours. The program also provides bird guide training and support for local projects to enhance birding sites for visitors.

BirdsCaribbean. 2023. "Endemic Birds: From the Nest." www.birdscaribbean.org/category/from-the-nest/

Library of educational resources available for free download on Caribbean endemic birds, including coloring pages, natural history information, photos, videos, puzzles, bird songs and calls, and activity sheets.

BirdsCaribbean. 2023. "Caribbean Endemic Bird Festival." www.birdscaribbean.org/caribbean-endemic-bird-festival/

Annual, month-long, region-wide festival to raise awareness of the Caribbean's endemic birds.

BirdsCaribbean. 2023. "World Migratory Bird Day." www.birdscaribbean.org/world-migratory-bird-day/

Day set aside globally for the recognition of migratory birds and their conservation, primarily in September–October but held flexibly in the Caribbean.

BirdsCaribbean. 2023. "Caribbean Endemic Species Checklist." https://docs.google.com/spreadsheets/d/1PAsnicPJYBLnISiKR2enQgJDAueoBuIHXGPv_Whisew/edit#gid=0

Online species list updated to latest taxonomic changes.

BirdLife International. 2023. "The Americas." www.birdlife.org/the-americas/

International bird conservation organization having 6 country partners in Caribbean and others along the Caribbean coast of Central and South America. Programs include Important Bird Areas, compiling IUCN RedList of threatened bird species, and flyway conservation initiatives as well as local bird conservation efforts by country partners.

eBird. date unknown. "About eBird." ebird.org/about

Global, participatory, online database of bird observations managed by the Cornell Laboratory of Ornithology. Anyone can contribute sightings, which then can be amalgamated to provide information about bird distribution and abundance. A way to record and share observations of birds, including in one's own garden.

Faaborg, J. 2002. *Saving Migrant Birds, Developing Strategies for the Future*. Austin: University of Texas Press.

Conservation issues facing migratory birds in the Americas.

Fitts, R. 2022. "Avian Superhighways: The Four Flyways of North America." American Bird Conservancy. abcbirds.org/blog/north-american-bird-flyways/#:~:text=Birds%20navigate%20along%20more%20or,%2C%20Central%2C%20and%20Pacific%20Flyways

An overview of the recognized flyways along which North American breeding birds move in their annual migration.

Gannon, M. R. 2005. *Bats of Puerto Rico: An Island Focus and a Caribbean Perspective*. Lubbock: Texas University Press.

Covers the bat species of Puerto Rico. Also, pertinent to the rest of the Caribbean.

Gerbracht, J., and A. Levesque. 2019. "Complete Checklist of the Birds of the West Indies." www.birdscaribbean.org/caribbean-birds

This site provides an updated list of bird species recognized from the Caribbean as well as background, definitions, and country distributions.

Hailey, A., B. S. Wilson and J. A. Horrocks, eds. 2011. *Conservation of Caribbean Island Herpetofaunas, Volumes 1 and 2*. Leiden: Brill.

Summaries and analyses of the species and regional conservation issues of Caribbean reptiles and amphibians.

Hallett, B. 2006. *Birds of the Bahamas and the Turks & Caicos Islands*. Oxford: Macmillan Caribbean.

Provides excellent photographs of Bahaman birds, helpful in identification of some of the birds found elsewhere in the Caribbean.

Kirwan, G. M., A. Levesque, M. Oberle and C. J. Sharpe. 2019. *Birds of the West Indies*. Lynx and BirdLife International Field Guides Collection. Barcelona: Lynx Editions.

A thoroughly illustrated guide also covering range, status, habitat, and behavior.

Kushlan, J. A., and K. Hines. 2014. *Attracting Birds to South Florida Gardens*. Gainesville: University Press of Florida.

Companion to the present book describing the birds and other wildlife that may be attracted to natural gardens and native plants in South Florida. Many plants and birds also occur in the Caribbean.

Mancina, C. A., and D. D. Cruz Flores, eds. 2017. *Diversidad Biológica de Cuba: Métodos de Inventario, Monitoreo, y Colecciones Biológicas*. Havana: Editorial AMA.

Multiauthor summary and synthesis of Cuban biodiversity incorporating analyses of animal and plant groups.

Merlin Bird ID. date unknown. merlin.allaboutbirds.org/

A bird identification App from the Cornell Laboratory of Ornithology with images, descriptions, range maps and sounds for birds from all around the world. Includes tools to help users identify birds both by sight and sound.

Navarro, N. 2015. *Endemic Birds of Cuba: A Comprehensive Field Guide, Including West Indian Endemics Residing in Cuba*. Crescent Beach: Ediciones Nuevos Mundos.

Available in English and Spanish, an illustrated guide focusing on endemic species. Provides species accounts along with background information about Cuba and its bird habitats.

Navarro, N. 2023. *Annotated Checklist of the Birds of Cuba,* 6th edition. Crescent Beach: Ediciones Nuevos Mundos. edicionesnuevosmundos.com/2023/02/12/annotated-checklist-2023/

Annually updated checklist of the birds of Cuba, keeping up with the fast-changing taxonomy of Caribbean birds. Printable or downloadable.

Raffaele, H. A., and J. W. Wiley. 2014. *Wildlife of the Caribbean*. Princeton: Princeton University Press.

An overview of plants and animals commonly encountered in the Caribbean.

Raffaele, H., J. Wiley, O. Garrido, A. Keith and J. Raffaele. 1998. *A Guide to the Birds of the West Indies*. Princeton: Princeton University Press.

Definitive work on birds occurring in the Caribbean as of publication date. Excellent illustrations help in identification, as does the comprehensive table of what birds are found on each island group at the time of publication. Each species is described as are useful details such as status, range, and habits. It is an indispensable book. An update of some information in more concise field-guide format is Raffaele et al. (2020).

Raffaele, H. A., J. Wiley, O. H. Garrido, A. Keith and J. I. Raffaele. 2020. *Birds of the West Indies*, 2nd edition. Princeton: Princeton University Press.

An illustrated guide to Caribbean birds.

Smith, D. S., L. D. Miller and J. Y. Miller. 1994. *The Butterflies of the West Indies and South Florida*. Oxford: Oxford University Press.

An illustrated guide to over 60 Caribbean species. Covers descriptions, distribution, habitat, and host plants. Expensive, probably mostly for libraries or devotees.

Turner, T. W., and V. A. Turland. 2017. *Discovering Jamaican Butterflies and their Relationships around the Caribbean*. Safety Harbor: Caribbean Wildlife Publications.

A comprehensive study of Jamaican butterflies and their conservation.

Wege, D. C., and V. Anadon-Irizarry. 2008. "Important Bird Areas in the Caribbean: Key Sites for Conservation." BirdLife International. datazone.birdlife.org/info/ibascarib

Provides information on the inventory of the Caribbean's Important Bird Areas.

8.3 Place-based conservation: protected areas, parks, and gardens

Acosta Cruz, M., L. Mugica Valdes and K. Aguilar Mugica. 2022. *Aves del Jardin Botanico Nacional de Cuba*. Havana: Sello Editorial AMA.

An excellent study of how birds use the National Botanic Garden of Cuba, which features areas of native plants arranged by habitat. Superbly illustrated, the book covers the birds, the plants, and their interactions.

Ahmad, N. 2011. *Soils of the Caribbean*. Kingston: Ian Randle.

A comprehensive overview and detailed analysis of Caribbean soils.

Alliance for Zero Extinction. 2023. "Global AZA Map." https://zeroextinction.org/site-identification/2023-global-aze-map/

Online map of Alliance for Zero Extinction sites accessible by site, country, or species.

BirdsCaribbean. 2023. "About Us." www.birdscaribbean.org/

Principal regional scientific and conservation organization on the forefront of conservation of native birds and their habitats. The website has comprehensive information. Programs include the *Journal of Caribbean Ornithology*, biennial International Conferences, Caribbean Waterbird Census, Seabird Conservation, Caribbean Landbird Monitoring (including PROALAS surveys, Caribbean Bird Banding Network and the Caribbean Motus Collaboration), Caribbean Birding Trail, and education programs including the Caribbean Endemic Bird Festival, World Migratory Bird Day, West Indian Whistling-Duck and Wetlands Conservation Project, and BirdSleuth Caribbean.

BirdsCaribbean. 2023. "Caribbean Birding Trail." caribbeanbirdingtrail.org/

A network of the best sites on each island for observing birds and otherwise enjoying nature. Includes information on featured birding sites, local guides, and tours. The program also provides bird guide training and support for local projects to enhance birding sites for visitors.

BirdLife International. 2023. "The Americas." www.birdlife.org/the-americas/

International bird conservation organization having 6 country partners in Caribbean and others along Caribbean coast of Central and South America. Programs include Important Bird Areas, compiling IUCN RedList of threatened bird species, and flyway conservation initiatives as well as local bird conservation efforts by country partners.

Bannochie, I., and M. Light. 1993. *Gardening in the Caribbean*. Oxford: MacMillan Caribbean.

Based on the long-term experience of a renowned Bajan garden expert and founder of the Andromeda Botanic Gardens, the book is chock-full of helpful hints on various bits of making a successful tropical garden. Like most such books of its era, its recommendations go to showy nonnatives, chemical control, fixing soil, and plant pampering. But offers practical advice the senior author accumulated over decades of engagement.

CANARI. 2019. *The Caribbean Islands Biodiversity Hotspot*. Final Version. Baratara: Caribbean Natural Resources Institute. www.cepf.net/sites/default/files/cepf-caribbean-islands-ecosystem-profile-summary-2020-english.pdf; CEPF. 2010. *Ecosystem Profile: The Caribbean Islands Biodiversity Hotspot*. Arlington: Critical Ecosystem Partnership Fund. www.cepf.net/sites/default/files/caribbean_ep_summary.pdf

Both these documents and related ones (www.cepf.net/our-work/biodiversity-hotspots/caribbean-islands) available through the Critical Ecosystem Partnership Fund website are the most complete, indispensable region-wide review of Caribbean biodiversity.

Coghlan, R. 1975. *Landscape Gardening in the Tropics*. London: Hart-Davis, MacGibbon.

Now quite old, written long before the modern natural gardening movement arose. That aside, provides excellent advice with special appreciation for special issues of the Caribbean. As the author's experience was in Barbados with traditional English garden, landscaping is the focus.

Collett, J., and P. Bowe. 1998. *Gardens of the Caribbean*. London: Macmillan Education.

This book has a couple of useful values. It provides appreciative glimpses of the great public and some private gardens of the Caribbean and a superb history of how Caribbean formal gardens came to be. These are seldom natural or naturalistic gardens, but the book captures well the historic European background of traditional Caribbean gardening.

Druse, K. 1989. *The Natural Garden*. New York: Clarkson N. Potter.

The premise of this book is close to that of the current book, although the author's concept of a natural garden is not a native plant garden. It is temperate and rather North American in approach. Its arguments and its photography are valuable.

Ferguson Fowler, C. H. 2013. *Backyard Gardening in the Bahamas and the Caribbean*. Published by author.

This instruction manual gets right to its advice as to what to do, covering a diversity of topics from the author's point of view expressed in simple terms. Anyone wanting an edible garden in the Caribbean can benefit from reviewing the author's instructions.

Ford, J. R. D., D. Ramroop, V. Boerger, V. Lopez, A. Pantoja, J. James and O. Elabanjo. 2013. *A Vegetable Garden for All,* 5th edition. Antigua and Barbuda: Zero Hunger Challenge. www.fao.org/docrep/019/i3556e/i3556e.pdf

An example of a backyard vegetable gardening manual, provided by an international organization, aimed at the family, children, and teachers, presented in simple terms.

Gann, G. D., T. McDonald, B. Walder, J. Aronson, C. R. Nelson, J. Jonson, J. G. Hallett, et al. 2019. "International Principles and Standards for the Practice of Ecological Restoration. Second Edition." *Restoration Ecology* S1–S46. https://doi.org/10.1111/rec.13035

Hall, S. 2011. *The New Caribbean Home Garden Handbook*. Trinidad: Guardian Media.

If you have one book on Caribbean edible gardening, this is it. Includes everything you might wish to know about vegetable gardening and more.

Kushlan, J. A., and K. Hines. 2014. *Attracting Birds to South Florida Gardens*. Gainesville: University Press of Florida.

Companion to the present book, providing details on the philosophy of natural gardening and how to create such gardens in South Florida. Pertinent also to the Caribbean, particularly limestone islands similar to South Florida.

Kushlan, J. A., and K. Hines. 2014. *Birds of Fairchild*. Morley: Acclaim Press.

A photographically illustrated study of how a public garden creates habitat for birds. Although set in South Florida, many of the plants and birds are found in the Caribbean and concepts apply.

Light, M. H. S. 1999. *Container Gardening for the Caribbean and the Tropics*. Oxford: Macmillan Caribbean.

Basics of container gardening that are universally applicable but, despite the title, not particularly focused on Caribbean uniqueness.

McGinley, K. 2023. "The Caribbean." In F. Cubbage, ed. *Forest and Forestry in the Americas: An Encyclopedia*. Society of American Foresters and International Society of Tropical Foresters. sites.google.com/site/forestryencyclopedia/Home

A short summary of Caribbean forestry.

Ranier, T., and C. West. 2015. *Planting in a Post-Wild World*. Portland: Timber Press.

Tries to fathom how humans are to persist in a world in which nature has been removed. This thoughtful discussion is a must-read in understanding how individual and community regreening decisions fit into a larger picture.

Reese-Warren, M. 2021. *The Ecological Gardener: How to Create Beauty and Biodiversity from the Soil Up*. White River Junction: Chelsea Green Publishing.

Focus is on the totality of gardening.

Robertson, W. 2011. *The Wild Garden*. Cambridge: Cambridge University Press.

First published in 1870, this seminal book challenged the formal garden, advocating a naturalistic approach.

Walters, B. B. 2019. *The Greening of Saint Lucia: Economic Development and Environmental Change in the Eastern Caribbean*. Kingston: University of West Indies Press.

An interesting study of changing agriculture and forestry practices in one of the most forested Caribbean nations. History applicable to the Caribbean more widely.

Weaner, L., and T. Christopher. 2016. *Garden Revolution. How Our Landscapes can be a Source of Environmental Change*. Portland: Timber Press.

Local recovery of nature using native habitats to inform landscape. It asks: "What plants want to be here?" There is not a better starting question for choosing plants for a natural Caribbean garden.

Winters, H. F., and G. W. Miskimen. 1967. *Vegetable Gardening in the Caribbean Area*. U.S. Department of Agriculture, Agriculture Handbook 323. Washington DC: Department of Agriculture. naldc.nal.usda.gov/download/CAT87209120/PDF

Ancient but classic text on Caribbean vegetable plants from the U.S. Department of Agriculture.

Woelfle-Erskine, C., and A. Uncapher. 2012. *Creating Rain Gardens*. Portland: Timber Press.

Aimed at temperate urban and suburban settings. Covers the many ways water can be harvested including design, construction, and uses. Provides technical details important in proper design.

Wege, D. C., and V. Ánadon-Irizarry. 2008. *Important Bird Areas in the Caribbean: Key Sites for Conservation*. Cambridge: BirdLife International. datazone.birdlife.org/info/ibascarib

Provides information on the inventory of the Caribbean's Important Bird Areas.

Wormser, O. 2022. *Lawns into Meadows. Second Edition: Growing a Regenerative Landscape*. San Francisco: Stone Pier Press.

Temperate-zone perspective in creating diverse lawns, including wildflowers and native grasses.

8.4 Conservation

American Bird Conservancy. 2023. "About ABC." abcbirds.org/about/

International bird conservation organization with programs including pesticides, feral cats, windows, and other threats to birds. Considerable information available on the website. Works with Caribbean partners on endangered species and habitat conservation. National partner of BirdLife International.

BirdLife International. 2023. "The Americas." www.birdlife.org/the-americas/

International bird conservation organization having 6 country partners in the Caribbean and others along Caribbean coast of Central and South America. Programs include Important Bird Areas, compiling IUCN RedList of threatened bird species, and flyway conservation initiatives as well as local bird conservation efforts by country partners.

BirdsCaribbean. 2023. "About Us." www.birdscaribbean.org/

Principal regional scientific and conservation organization on the forefront of conservation of birds and their habitats.

Burbidge, B., and P. S. Wyse Jackson, eds. 1998. *Conservation Action Plan for Botanic Gardens of the Caribbean Islands*. London: Botanic Gardens Conservation International. www.researchgate.net/publication/261213798

An analysis and plan for increasing the engagement of Caribbean botanical gardens in conservation.

CANARI. 2019. *The Caribbean Islands Biodiversity Hotspot*. Final Version. Baratara: Caribbean Natural Resources Institute. www.cepf.net/sites/default/files/cepf-caribbean-islands-ecosystem-profile-summary-2020-english.pdf; CEPF. 2010. *Ecosystem Profile: The Caribbean Islands Biodiversity Hotspot*. Arlington: Critical Ecosystem Partnership Fund. www.cepf.net/sites/default/files/caribbean_ep_summary.pdf

Both of these documents and related ones (www.cepf.net/our-work/biodiversity-hotspots/caribbean-islands) available through the Critical Ecosystem Partnership Fund website are the most complete, indispensable region-wide review of Caribbean biodiversity.

Gann, G. D., T. McDonald, B. Walder, J. Aronson, C. R. Nelson, J. Jonson, J. G. Hallett, et al. 2019. "International Principles and Standards for the Practice of Ecological Restoration. Second Edition." *Restoration Ecology* S1–S46. https://doi.org/10.1111/rec.13035

Matloff, G., C. Bangs and L. Johnson. 2010. *Paradise Regained: The Regreening of Earth*. New York: Copernicus Books.

Discusses how civilization has changed natural ecological systems and potential for restoration and regreening.

Maunder, M., A. Leiva, E. Santiago-Valentin, D. R. Stevenson, P. Acevedo-Rodriguez, A. W. Meerow, M. Mejia, et al. 2008. "Plant Conservation in the Caribbean Island Biodiversity Hotspot." *The Botanical Review* 74:197–207. www.researchgate.net/publication/226784657_Plant_Conservation_in_the_Caribbean_Island_Biodiversity_Hotspot

A review of the status of plant conservation in the Caribbean with specific recommendations for improving plant conservation in the Caribbean.

Mittermeier, R. A., P. R. Gil, M. Hoffman, J. Pilgrim, T. Brooks, C. G. Mittermeier, J. Lamoreux and G. A. B. da Fonseca. 2005. *Hotspots Revisited: Earth's Biologically Richest and Most Endangered Terrestrial Ecoregions*. Chicago: University of Chicago Press.
A review of biodiversity hotspots.

Richards, M. A. 2018. *Regreening the Built Environment: Nature, Green Space, and Sustainability*. New York: Routledge.
Discusses interrelationships between nature and the built environment.

United Nations Environment Programme. 2004. *Caribbean Environment Outlook*. New York: UNEP. wedocs.unep.org/handle/20.500.11822/9376
An analysis and summary of the state of the environment in the Caribbean providing background on the major issues of environmental degradation and reduction in natural services provided by the environment.

Wege, D. C., and V. Anadon-Irizarry. 2008. *Important Bird Areas in the Caribbean: Key Sites for Conservation*. Cambridge: BirdLife International. datazone.birdlife.org/info/ibascarib
Provides information on the most important sites for birds in the Caribbean, including what birds to expect there.

Index

Note: Page numbers in *italics* indicate a figure on the corresponding page.

C

D

Q

R